Ephraim Cutter

A Contribution to the Treatment of Uterine Versions and Flexions

Second Edition

Ephraim Cutter

A Contribution to the Treatment of Uterine Versions and Flexions
Second Edition

ISBN/EAN: 9783337222802

Printed in Europe, USA, Canada, Australia, Japan

Cover: Foto ©berggeist007 / pixelio.de

More available books at **www.hansebooks.com**

A Contribution

TO THE TREATMENT OF

Uterine Versions and Flexions.

BY

EPHRAIM CUTTER, A.M., M.D.

SECOND EDITION, ENTIRELY REWRITTEN.

BOSTON:

JAMES CAMPBELL, PUBLISHER.

1876.

Cambridge:
Press of John Wilson & Son.

TO

T. GAILLARD THOMAS, M.D.,

The Eminent Gynæcologist,

THIS LITTLE WORK IS DEDICATED, BY PERMISSION.

PREFACE

TO THE SECOND EDITION.

A S said in the first edition, this is not intended for a piece of fine writing. It is simply a contribution from one who believes it the duty of every one to strive to leave his profession better than he found it. The first edition being out of print, a new one is called for. The instruments herein described have been commended, and widely adopted. It is thought to be no more than right that those who use the writer's supporters should have a clear conception of the inventor's ideas in their employment. I have been pained to learn that some instrument makers have produced my pessaries, or what they call my pessaries, and have left out the very feature I most value. A patent would have protected the instrument; as it is, I must rest the matter simply upon a respectful protest against this naming of bastards after me.

The writer has been permitted to know that the
devices suggested here have relieved large num-
bers of suffering women. This knowledge is an
incentive to this renewed attempt to introduce this
set of pessaries to a larger acquaintance in the profession. Some may find fault with the style ; but
the *things* stated must not be blamed or discarded
because the inapt manner of their presentation
may be open to just criticism.

It may be added that CODMAN & SHURTLEFF
make my instruments according to my models and
satisfaction.

E. C.

13 TEMPLE STREET,
  July, 24, 1876.

CONTENTS.

LIST OF ILLUSTRATIONS.

———◆———

INTRODUCTION.

In order to understand the principles of the proposed treatment, attention is solicited to a very brief enumeration of the important features which obtain when the uterus is *in situ naturali et sano*.

Fig. 1 represents in section the normal position of the womb. The vagina, — a muscular membranous tube, open at the lower end, curved antero-posteriorly, its long axis running upwards and forwards, and the antero-posterior wall nearly in contact (the artist has failed to show this contact), — occupies the central portion of the picture. At its upper part lies the uterus, inclined somewhat forward and projecting into the vagina, which unites with the outer walls of the uterus in such a manner as to form a pocket (*cul-de-sac*) or sulcus round the neck of the womb. In the figure, the uterus appears in section, like a horseshoe magnet shut closely. Owing to the flattened condition of the uterus antero-posteriorly, the *cul-de-sac* is divided into anterior and posterior (Douglass) for practical purposes. Sometimes there is no *cul-de-sac;* the uterus just projecting into the vagina, but not enough to form a recess. In these cases, dif-

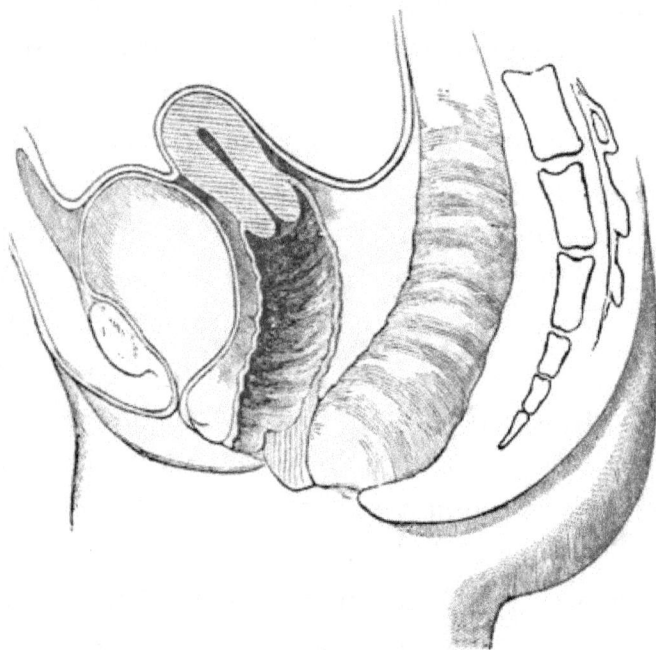

Fig. 1.

ferent physical arrangements must be made for the correction of the displacements than where the *cul-de-sac* exists. The broad and round ligaments are important features of support. It would seem that they of themselves should be sufficient to maintain the uterus in place. In front are the intestines, bladder, pubis, and investments. Behind are intestines, sacrum, and coverings. The uterus sits on an elastic support, the vagina, stayed and guyed by the broad and round ligaments, and bolstered by intestines. As long as the vagina (and this is the old doctrine) maintains its normal character, — that is, has its full length extended in natural antero-posterior curve, has its antero-posterior surfaces lightly in contact, and the transverse fibres of the vagina normally contracted, forming a continuous sphincter muscle from vulva to uterus, — it is almost impossible for a displacement of the normal-sized uterus to occur. To repeat : — the tonicity of the transverse muscular fibres of the vagina (sphincters) causes the natural axes and curves to be preserved, and holds up the womb ; just as the sphincter ani muscle does the rectum and its contents. The writer is aware that this doctrine is not received by Thomas, and perhaps others ; still, with all due deference, the position is here respectfully maintained, that, as in complete laceration and separation of the sphincter ani muscle, there is a want of normal retention of the natural contents of the rectum, — so, in the dilatation of

the sphincter vaginal muscle (not a short affair
of a half-inch, but of three or four inches), it
must be expected to find that there will be a
dislocation of the normal occupant of this extremity. Indeed, what more physically perfect
support can be found than the normal vagina
presents? — how very analogous to an elastic and
yielding spiral spring, which presents the most
admirable features for keeping up the womb! In
healthy condition, it cannot go below the perineum ;
and of course the transverse contraction extends,
or, as it were, erects it in the other direction, just
where it is needed to hold the uterus in place during the functions or other movements of the contents of the abdomen. This subject is one of so
great importance, that further attention is solicited
for it. Leaving out of consideration the uterus
for the present, let us study the possible deviation of the vagina. Recalling that it is a tube,
three to four inches long, flattened antero-posteriorly, with its front and back parietes slightly
in contact, curved also anteriorly, lying loosely
in the roomy cavity of the pelvis, and one end
resting on the perineum, — let us inquire what
its possible displacements may be. They may

be (a) upwards ; (b) centripetal ; (c) centrifugal ; (d) downwards.

(a) *Upwards.* — It is difficult to conceive that the normal vagina, contracted normally and extended in its normal length, can be displaced upwards, from the simple fact that its vulval end is attached to the coccyx. Hence, we dismiss the possibility of an upward displacement.

(b) *Centripetal.* — By this is meant the displacement of the vagina towards its own central axis. This it is. also difficult to conceive of, as, if such a thing were possible, it would only the more erect or stiffen the vagina by bringing its walls more closely in contact.

(c) *Centrifugal.* — This includes all lateral or sidewise displacement. The vagina cannot be moved forwards without infringing upon and being arrested by the concave surface of the os pubes and the vesical apparatus. It cannot go backwards, because it is fastened by its concave antero curve or face to the pubic bone and bladder. There may be a thinning of the vaginal wall, and a distension from within by gasses or atmospheric pressure, — as when the trunk of the patient is depressed below the level of the pelvis, and the viscera sag down : air then will penetrate the vagina and distend it like a ball ; but this condition disappears when the erect position of the trunk is maintained, and is only temporary. Indeed, it has been suggested to make use of this distension to right abnormal vaginæ ; but its fallacy lies in the fact that this process abnormally distends the vagina.

(d) *Downwards.* — This is possible only when the normal tone of the transverse muscular fibres is weakened or destroyed ; for, as we have already seen, the vagina cannot be bent because of its intimate connection with the os pubis. The vagina, then, being weakened, it is readily seen how it may be " squashed," shortened transversely, wrinkled, and folded : the antero-posterior walls are longer in coming into accurate apposition, but touching, it may be, at the tips of the transverse wrinkles. It is the posterior wall that wrinkles the most, as the anterior is held by the pubic attachment so often referred to. But, to repeat, it is impossible to conceive of the posterior wall wrinkling transversely, if those transverse muscular fibres (sphincters) are normally tonically contracted. So that we are compelled to infer that a normal vagina cannot be displaced without destroying its normality. Given, then, a *normal vagina*, how can the uterus be displaced ? It is evident that the uterus must be displaced upwards, downwards, forwards, or backwards.

Upwards. — It cannot go upwards, as the vagina holds it ; and besides, it has to encounter the weight of the intestines and pressure from the diaphragm.

Downwards. — The vagina being, by the terms of our simile, in a normal condition of tonicity, the contraction of the continuous sphincter of the vagina would not allow the descent of the uterus into the grasp of those very transverse muscular fibres.

Forwards. — If the uterus tips forwards on the end of the normally-contracted and extended vagina, it must stretch the back section of the upper part of the vagina, and expand the front section of the same, or invert it — turn it in so that the anterior utero-vaginal *cul-de-sac* is obliterated. But these processes involve a disturbance of the normal condition of those transverse fibres, and also a departure from the terms of our proposition ; namely, that of a normal vagina, and a displacement of the womb. Throwing aside all this, there is but little chance for the uterus to move forwards over the end of a normally-contracted vagina, on account of

the little room or space.   Naturally the uterine axis
is inclined at a considerable degree forwards, so that
an increase of this forward obliquity at this normal
elevation would bring it into contact with the
bladder and abdominal walls in the hypogastrium.
The effect of the broad ligaments is to prevent
forward displacement.

Backwards. — The same physical changes as just
described in the last section would obtain in a backward version of the uterus, — only reversed.   The
anterior insertion of the vagina would be stretched
and pulled upwards, and the posterior section flattened and folded against the posterior wall of the
uterus.   It is not intended to say that such a condition of things is impossible, but it is desired to
state that before the posterior section of the vagina — acting as a fulcrum — allowed the uterus
to turn back, it would be "squashed" down and
give way, and thus become an abnormal vagina.
The effect of the round ligaments would also be
to keep the uterus from going backwards.

We therefore respectfully adhere to the old doctrine of normal vagina, and normally-posed uterus;
including in this proposition the normality of the
uterine ligaments.   For without these natural guys

and supports, the vagina cannot long maintain its tonicity. We may here add, that, in our own experience and personal knowledge, we have yet to find a normal vagina and a displaced uterus. We have seen one case of ante-version, where the vulval half of the vagina was normally contracted, and the walls properly in contact; but the uterine or upper half of the vagina was dilated and expanded laterally, — the posterior wall touching the os uteri, and the anterior the fundus uteri; so that the top or dome of the vagina was all womb.

It is quite evident that ordinarily we can directly do little for the restoration of the relaxed uterine ligaments, as we cannot get at them. The round ligament might be exposed by abdominal section shortened and secured; but one of our greatest operators in abdominal surgery (Dr. Kimball) says it would be entirely unjustifiable. We are, then, confined to efforts which will maintain the normality of the vagina, sustain the uterus in place, and thereby relieve the mal-condition of the ligaments. It is legitimate to suppose that, having thus put Nature in a way to help herself, absorption and tonicity will return in time to the stretched ligament; as in the case of fractured limbs, the surgeon only puts the injured parts into the best possible condition for the ·healing processes. The principles of treatment, then, may be forestated to be : —

1. *Uterine and vaginal restoration by the uterine sound.*

2. *Maintenance there by means mechanically adapted to the parts ; which allow of the normal contraction of the transverse vaginal muscular fibres, and which permit a natural degree of uterine mobility.*

RETROVERSION.

Retroversion of the Uterus is that condition of the womb in which it is bodily displaced backwards, while the long axis is not bent on itself, but pre-

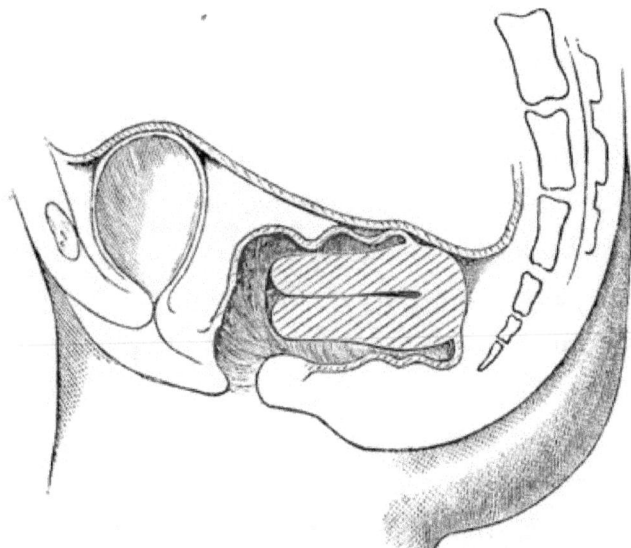

Fig. 2.

serves a rectilinear direction. The amount of backward obliquity varies from a displacement of a few degrees from the natural position to almost one hundred and eighty degrees. Taking an instance, where the variation is about one hundred and thirty-five degrees (a common case), — we find the fundus pointing towards the concavity of the sacrum, and the os is directed towards the pubis. When the forefinger is passed into it, it finds the vagina shortened in its long axis and lengthened in its transverse diameter, and it comes directly in contact with the posterior surface of the womb. (*Vide* Fig. 2.) The posterior *cul-de-sac* is not well defined; the posterior vaginal wall being reflected back, and lying in contact with the womb. In one case, the vagina seemed to have been, as it were, peeled from the uterus and drawn upwards, so that the posterior *cul-de-sac* was found at the fundus. To be able to introduce the uterine sound for diagnosis, the os must be drawn downwards by the fingers, the concavity of the sound turned backwards, while the handle protrudes forward between the thighs. Generally, the uterine cavity is found to measure more than two and a half inches, and the uterus congested and hardened. The exploration then reveals, —

1. A vagina shortened in its long normal axis.

2. A vagina lengthened in its short normal axis, thus reversing the natural order of the axes ;

namely, the short diameter becomes the long, and the long diameter becomes the short. *This feature constitutes the essential factor of the unnatural condition.*

3. Posterior uterine wall forming the dome of the vagina.

4. Fundus felt towards sacrum, os towards pubis.

5. Sound passing in with concavity backwards, handle in front.

6. A settling down of the whole womb into the pelvic cavity bodily, and yet at the same time intra-vaginal mostly, not extra-vaginal.

Ætiology. — The ætiology of retroversion does not belong to this portion of this work. A large variety of causes have been enumerated, any or all of which may have operated to produce the result. It is sufficient to remark that they must have acted on mechanical principles, either by overwhelming the broad and round ligaments and the vagina, or weakening those supports, so that, when the organ is subjected to the motions which physiologically or accidentally occur, they give way. There must then result a relaxed and shortened and distended vagina, and a relaxing and lengthening of the uterine ligaments. A case of retroversion being clearly

made out, and not confounded with pregnancy or
any other diseased or pathological condition, the
indications for treatment may be described as
follows : —

1. As already pointed out, the uterine sound
should restore the uterus and vagina to their normal position.

2. They should be maintained there by means
mechanically adapted to the parts, which allow of the
normal contraction of the transverse vaginal fibres,
and which permit a natural degree of mobility.

3. Use unirritating material.

4. Instruments employed should be manageable
by the patient.

5. Pay attention to the general health of the
patient.

In the instruments to be described, the writer
has endeavored to carry out these principles. If
he has not entirely succeeded, it is hoped that a clew
may be given for others to follow out to perfection
in this difficult department of mechanical appliances
to the human body. Two forms of retroversion
pessary are submitted, — the Loop and the T.

The Loop Pessary of the Writer. — This includes
a belt of inelastic webbing, thirty-six inches long,

and one and one-fourth of an inch wide, to go squarely around the waist ; and a suspensory cord of india-rubber tubing, ten inches long and one-fourth of an inch in diameter, attached to the middle of the belt by a loop and cord. This is designed to run through the natal furrow (between the buttocks), and is also attached to the perineal extremity of the pessary. It insures an elastic support, and the furrow prevents lateral motion.

The pessary is a cylinder of hard rubber, about one-fourth of an inch in diameter, curved into a hook one and one-fourth of an inch to two and one-half inches in diameter, intended to go around and entirely clear the perineum. The perineal hook of the pessary should not approach the perineum nearer than half an inch. The pessary then curves backward and divides into two branches, curved to correspond with the vaginal curve and terminate by uniting together into a loop. This loop is turned backwards so as to fit into the post utero-vaginal *cul-de-sac*, and allow the cervix of the womb to project into the fenestra made by the loop.

In the middle of the hook there is a joint, so that it may be turned out of the way during defecation, and (what is of very great importance) allow of a

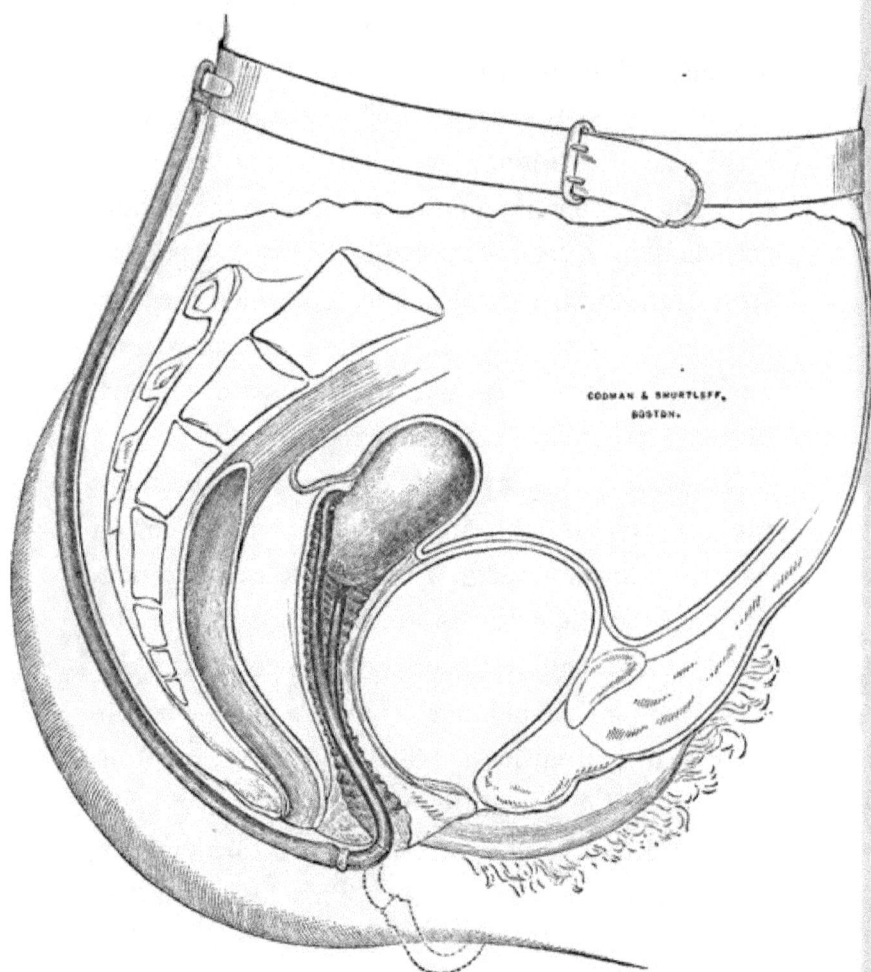

Fig. 3. — Vertical section of pelvis, showing retroversion pessary applied.

support during this act, where the bearing down efforts have a direct tendency to produce retroversion again, and will certainly cause it, if not prevented, in most instances. The design of this mechanical contrivance is accurately to fit the flat antero-posterior curve of the vagina. The loop is engaged in the post-uterine vaginal *cul-de-sac*, and thus holds the uterus in place ; extends the vagina in its long diameter without unduly distending the transverse vaginal fibres, or preventing the normal tonic contraction of these very fibres, — which, as we have already seen, is a most important factor in a healthy vagina. It is not intended to have the instrument take any bearing upon the lateral vaginal walls, for the same reason that it is not common-sense to distend still more the already abnormally distended and relaxed walls. The posterior portion of the pessary, for the same reason, should not press on the *perineum.* Besides, *irritation* should be avoided by undue contact. The part which projects over the perineum backwards should take the natural curve of the vulval furrow in continuity with that of the natal furrow. This part, also, should not touch the anus. If it does, the curve is wrong, and the pessary should not be used. As a general

2

rule, the larger the curve of the pessary the better. Indeed, the enlargement of the perineal curve is about all the new improvement that time and use have suggested since the introduction of the pessary.

" In some cases, so very great is the pressure excited by the displaced uterus, that no purely internal support will answer," . . . as ulceration will result. " Under these circumstances, I have obtained the most gratifying results from the use of a modification of Cutter's retroversion pessary, intended to obviate a difficulty which I found to attend that excellent instrument, — that of cutting through the vagina. For this reason, I have affixed to the top of Cutter's pessary balls of different sizes, — some as large as a hickory-nut, — for the object is not only to prevent cutting of the vagina, but to place behind the displaced fundus a mass which will make it fall forward by *displacement*, not pressure. My alteration of this instrument is insignificant ; the entire credit of it belongs to Dr. Cutter, to whom I feel personally indebted for affording me so valuable and simple a method for meeting the difficulties of aggravated retroversion. Had I space, I could cite a number of very bad cases of this difficulty, which for years had resisted treatment by ordinary pessaries, and which have readily yielded to the use of these modified pessaries. . . . It is a painless and efficient means of giving support, and will gain a high reputation on account of these qualities in posterior displacements. The class of cases to which it is especially applicable is that in which the displacement is due to a prolapse of the posterior vaginal wall from rupture of the perineum or other cause. . . . This instrument should be removed every night, and replaced every morning. It may be said that this will prove difficult of accomplishment for the patient. Out of several hundred cases in which I have used it, I have never found an instance of failure in this respect. The patient will very often become disaffected towards the instrument from its chafing the perineum. By a little patience, covering the points that rub with greased lint, and leaving the pessary out until the irritated part be healed, the feeling will soon pass away." *

The writer would remark, that the perineal curve must have been too small in these cases. The trouble would best be obviated by a large two-and-

* Thomas's " Diseases of Women." Fourth edition. pp. 385 and 386. Philadelphia : 1874.

a-half-inch hook.   In my own practice, the perineum rarely chafes.

The T Pessary. — This is like the loop pessary,
except in the vaginal portion.   Instead of dividing

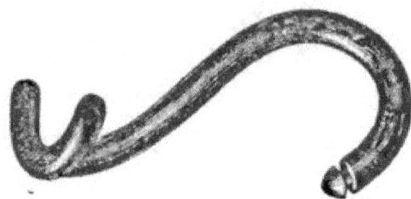

Fig. 4.

into two branches, the single cylinder is continued
upwards, with an antero-posterior curve, till it
reaches the post-vaginal *cul-de-sac* ; here it is terminated by a bar like that of the letter T : hence
its name.   This bar is a cylinder of the same
diameter as the stem, — three-quarters of an inch
in length or more, — curved forward, so as to form
a section of a circle the diameter of which is that
of the outside of the uterine neck.   In this manner,
its anterior concavity fits the convexity of the uterine wall, being intended to give an accurate fit to the
cul-de-sac.   The curve of this pessary is more analogous to that of the vagina than the loop pessary.
It is lighter and simpler, but more difficult of intro-

duction. If this objection were removed, it would take the precedence. The following represents a device to obviate this objection. The curve is wrong, as the artist has given an anteversion pessary.

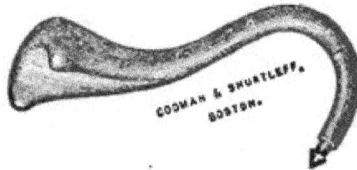

Fig. 5.

Application of the Loop Pessary. — The patient should be placed on the left side, lying upon a table or bed, or other support. The left arm may be projected out behind, bringing the left mamma in contact with the supporting surface. The knees should be drawn up towards the chin. In extraordinary cases, the prone position on the chest and knees may be adopted. The forefinger (anointed with soap and water) of the operator's left hand is introduced up to the os uteri. On this finger is passed the uterine sound, convexity forwards, into the uterus, as before described. The uterus is reinstated by turning the sound in the opposite direction, convexity forwards. This is a procedure to be done with care. It may be best accomplished by steadying the sound at the os uteri, and gently sweeping it about, describing a large semicircle by the handle; then, drawing the handle backwards, elevate the womb into its natural position. In doing this, a knowledge of the condition of the womb itself, and the surrounding parts, will be acquired. If there is no other lesion but the retroversion, there will not be much difficulty in accomplishing the task. If there is difficulty, then the attempt to adapt a pessary should be abandoned until that difficulty be removed. Of this, more may be said farther on. If the obliquity of the retroversion is small, the reinstation by the sound may be omitted, and the pessary fitted without.

Vaginometry. — Formerly, I was in the habit of roughly measuring the distance from the perineum to the posterior utero-vaginal *cul-de-sac*, by the finger, and then, selecting a pessary corresponding to the rude measurement as near as possible, apply it; and, if it did not fit, taking a number of pessaries, keep trying till a fit was procured. However, the finger was found unreliable, as it has no graduation marks; besides being jointed, moving involuntarily, and often too short. Moreover, any mechanic would laugh at such a measuring instrument in his arts. To remove this opprobrium from such important procedures as the present, involving the interests and welfare of a sex to whom all are indebted for existence, comfort, and enjoyment, the writer was led to institute a process which he has denominated *Vaginometry;* and by which accurate measurements may be obtained of the anterior and posterior walls of the vagina, and also of the diameter of the uterus at the point of projection into the vagina. This procedure seems to me to coincide with the dictates of common-sense ; as no artist in the various departments of appliances to the body, for the sake of protection from the inclemency of the weather, will undertake to make the products of his art without some system of measurement which is accurate and reliable. The instruments devised are two in number, — the vaginal sound, and the vaginometer.

Vaginal Sound. — This consists of a stout copper wire, nickel plated, eight

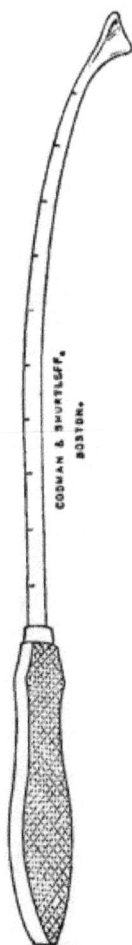

Fig. 6.

inches in length, graduated to half-inches, and
bent to correspond with the vaginal curve.   At the
proximal end is a handle, serrated on the front.
At the distal end, another piece of like wire, three-
fourths of an inch in length, and curved to an arc
of a circle three-fourths of an inch radius.   It is
very much like the uterine sound, with a cross-bar.
The point of an ordinary sound is too pointed, and
would not give an accurate result, because of pushing in beyond where it was intended.

The method of employment is seen in Fig. 7,

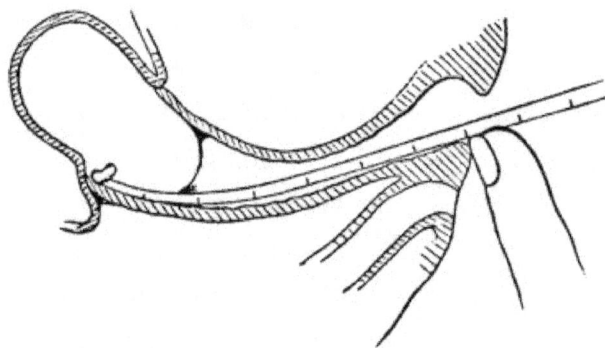

Fig. 7.

and consists in having the uterus held *in situ* by
the uterine sound (the sound is not shown in the
cut), the passing the left forefinger up to the os, the
placing the concavity of bar on convexity of finger,

and sliding the vaginal sound up into the *cul-de-sac.*
The point of emergence from the vagina is noted
by the forefinger of the right hand. ˊ The sound is
withdrawn as it was introduced, — gently moving it
outwards, as the ends of the bars are apt to catch
in the folds of the mucous membrane, and hurt the
patient unnecessarily.*   All this time the right
forefinger is kept at the point of the sound where
it was first placed, and an accurate admeasure-
ment is procured.

The Vaginometer is a combination of two vaginal
sounds, handled and jointed together at the proxi-

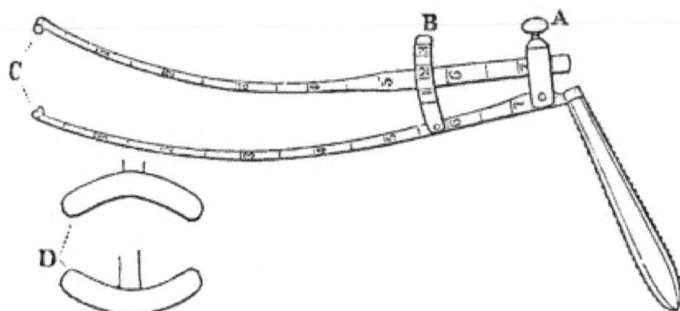

Fig. 8.

mal ends.   At the distal ends, the concavities of
the bars look towards each other.   Near the joint
is a graduated scale which gives in inches and
fractions the distance between the bars, not the

* Fig. 6 shows the angles of the bar and staff filled.

actual distance on the scale. The mode of application is: preparation as before described in the
use of the vaginal sound; to close the bars over
the uterine sound, and press up into the vagina,
so that the forward bar rests in the anterior *cul-de-
sac*, and the backward bar in the posterior *cul-de-
sac*. The graduated scale will give the diameter of
the uterine neck, and both anterior and posterior
vaginal walls are measured at once. When the
vaginometer is used, one is enabled to get the
radius of the curve of the inside of the loop, or of
the bar of the T, — a matter of some importance
in sensitive cases. It also furnishes the only
method known to the writer of obtaining a measurement for a cup pessary in prolapsus.

We now return to the application of the retroversion pessary. Having determined the length
of the posterior vaginal wall, and ascertained the
size of the uterus by the vaginometer, we select
a pessary one-half inch longer than the measurement, and with a curve in the loop inside corresponding to that of the cervix, and of course accurately fitting it. If the pessaries on hand do not
correspond, one may be ordered of the instrument-
makers, of whom Codman & Shurtleff are the best

acquainted with the instrument, and who make any desired size at ordinary rates.

Having, then, a pessary thus selected, — still maintaining the uterus *in situ* by means of the sound, as before described, — the loop pessary is passed over the perineum, behind the sound into the vagina, and gently pushed in until the loop is engaged in the posterior utero-vaginal *cul-de-sac.* If the measurements have been accurately made, and the, pessary corresponds to them, a perfect and easy fit will be found to have been accomplished. It is well to pass up the forefinger, and verify the accuracy of the placement. The sound is then withdrawn, and the uterus is held in its place by the pessary, the vagina is extended to its normal length (or as far as it is proper under the circumstances), the vaginal transverse fibres and the uterine ligaments are allowed an opportunity to contract, and the bends and curves of the pessary accurately to fit those of the vagina and posterior *cul-de-sac.* The perineal hook should surround the perineum without touching. If air does not freely circulate between it and the perineum, a longer pessary should take the place. The point of the hook should lie parallel with the curve of the natal

and vulval furrow, and should not impinge anywhere. If it does, a larger curve shóuld be employed. We would reiterate here that the curves
cannot well be too large. From the time of just
before withdrawing the uterine sound, a hold is
being kept on the suspensory cord, and it is passed
up into the natal furrow, and not relaxed until the
belt is squarely secured around the waist.

Tension of the Tubing Suspension.—This should
be only tight enough to be comfortable. It should
not be slack nor taut. Frequently the suspension
is too long for small females. In these cases, the
connection of the cord with the belt should be
severed, and the cord drawn up till the slack is
taken in. A little twist of the cord itself in a clove-
hitch is all I generally use. As a general thing, it
is advisable to have the cord rather tight at the
outset, as it can readily be controlled by relaxing the belt, and because the tubing is apt
to stretch. The quality of the tubing is an
important element of success. The instrument-
makers named have been very particular about
this detail. The constant use of soft rubber
causes an interstitial change of molecular state,
that seems to consist of a natural crystallizing

out of the substance, thereby impairing the elasticity. Indeed, these changes occur in time at best, but are deferred the longer according to purity of the rubber.

Before defecation, the patient should be instructed to loosen the belt about two inches ; then, turning the hook on its axis, to hold it and the tubing out of the way, supporting also the uterus during the bearing-down efforts of this function, and secure it at the close of the operation by tightening the belt. This procedure is one of the most difficult parts of the wearing the supporter ; but it is the most satisfactory method of preventing a displacement, which is almost sure to occur without some such outside support. At no other time during the wearing of this instrument is the womb liable to lapse, if the proper tension is steadily kept upon the suspensory cord. If the displacement is not large, the pessary may be entirely removed during defecation, washed, and replaced by the patient herself afterwards. Judgment, however, must be used in this respect.

For the T Pessary, the mode of introduction and management is the same, except that it can be introduced without the uterine sound, though better with.

One arm of the bar is placed in the vagina, inside the perineum ; then pushing in and twisting, the other arm is turned in, and the bar slid along the posterior vaginal wall into place. This pessary is purely my own. Of the loop pessary, the vaginal portion was adopted from a supporter of London make, which was invented and designed for the treatment of rectocele. The backward bend of the loop, the perineal curve, the joint, the single posterior elastic suspension running through the natal furrow, are features added by the writer. In some cases, I have found that the loop sometimes slips about sideways, in large uncontractile vaginas, causing lateral displacement of the instrument, and, of course, discomfort. To avoid this, I devised the T pessary.

It fits more accurately in the *cul-de-sac.*

It is wider on the top laterally.

A sidewise force has only the end of the bar to operate upon, whereas in the loop there is quite a long surface on the side to press against ; hence, it cannot tip so readily. The lateral distension of the vagina is reduced to the minimum, as the cylindrical stem is so narrow and occupies so little room as compared with the loop.

It fulfils perfectly the indications of a pessary, which (to repeat) are, —

(*a*) Extension of the vagina to normal length in the direction of the normal long vaginal axis.

(*b*) No distension laterally of the transverse vaginal muscular fibres, or, more properly, the vaginal sphincter. This is a fault of all the intra-vaginal pessaries so much employed at present by the profession at large. It seems as if the lateral distension of the diseased conditions was already sufficient, without making it any more. This criticism applies to the sponge pessaries, the soft-rubber pessaries, the globe pessaries, — the distension of the vagina by atmospheric pressure process, — indeed, to all the means whereby a support for the dislocated uterus is supplied by resting upon the lateral walls of the vagina, anteriorly, posteriorly, or against the pubis. The indication is to restore the normal tone to the vaginal sphincter muscle; but how can this be done by means which act directly to *stretch it still more?* The T pessary offers no lateral resistance to the sphincter, because it is so small as not to stretch it.

(*c*) Support of the retroverted womb by coming up *behind* it, and not allowing it to sink back any

further. If the support be given in retroversion
by means of a cup, the uterus may slip and turn
backwards over the edge of the cup. Of course, it
would be idle to think of supporting this form of
version by building up in the anterior *cul-de-sac*, as
this would tip the womb over back also ; so that
practice and common-sense both indicate the necessity of having the point of support of a retroverted womb in the post-utero vaginal *cul-de-sac*.
This would be superfluous explanation, were it not
that this principle is ignored "right straight along"
by persons treating these displacements. And is
it any thing strange that mechanical treatment by
pessaries should have fallen into great disrepute
and general condemnation, when the very simplest
physical laws of mechanical support are ignored
and disregarded ?

(*d*) Less liability to get out of place. When the
suspension of the T pessary is kept in the line of
the natal furrow, it lies in the median line of the
vagina vertically. It is difficult to conceive how
the pessary may be displaced from below, as the
natal furrow prevents lateral motion. If the pessary is disturbed, it must be, then, at the upper
end. Here the forces of displacement must act

downwards, but the elasticity of the suspension
will restore the T pessary to its place ; so that this
amounts to nothing. Then the forces of displacement must be lateral ; but we have already seen
how small a surface these side-thrusts must have
to act upon, and hence how little of power. It is
in those cases of very large, baggy vaginæ, where
there is no place for the ends of the bar of the T
to rest against, that the T can slip sideways ; and,
to go far, one arm of the T must rise and the other
descend. But, to do this, there will have to be a
lifting against the upper surface of the rising arm,
increasing as it rises. Now, if the vagina is
extended to its normal length, — unless there is
parietal thinning, — the tendency of the T pessary
is to straighten out these lateral wrinkles, or baggi-
ness, and transfer them into longitudinal wrinkles,
where they ought to be. These efforts are all in
the direction of cure. The lateral distention of the
vagina must govern the size of the bar. It seems
to the writer, after patient criticism, that the T pessary is therefore correct in its principles and con-
struction ; practically, however, I have found some
patients wearing the loop pessary with comfort,
who disliked the feeling of the T, and *vice versa.*

One Pessary not always Sufficient in each Case.
— Take an instance where the disease has existed
for a long time. The uterus has been elevated into
position. The distance from the posterior fourchette and the posterior *cul-de-sac* has been measured, and a pessary fitted. The vagina, relieved
from its contracted position, will relax and become
elongated more than at first. There is, then, some
discomfort and irksomeness. It now becomes
necessary to remove the pessary, pass the sound
into the uterus, reinstate the latter, and then measure over anew. It will be found, for instance, that
the distance has increased by half an inch or more.
It is only necessary to apply a correspondingly long
pessary. Should this become uncomfortable, repeat the process, and so on until the vagina reaches
its normal length. In these procedures, one may
begin with a very short instrument, and end with
quite a long one.

Third Indication. — To have unirritating material of which the pessaries are formed. Hard rubber, according to the experience of disinterested
parties, has been found almost unexceptionable.
Its use in contact with mucus membranes is more
extensive than any other material, — as witnessed

in dentistry. Its cheapness, lightness, strength, exquisite polish, and plasticity under moderate heat are unexampled combinations of excellences. It does not absorb or retain the vaginal fluids ; indeed, nothing better can be asked for. Silver plated with gold is a model substance for the vaginal use, but its expense is an objection.*

Fourth Indication. — To have the instrument manageable by the patient, to some extent at least. When the uterus is properly reinstated, and the pessary rightly and well applied, there is almost always an expression of relief on the part of the patient. If this is not the case, something is wrong either with the instrument, vagina, or womb. *In all* cases, if there is trouble and inconvenience, she is instructed to withdraw the instrument. Thus any mischief may be prevented. The physical form of the loop pessary renders it very easily removed by simply pulling it out by the perineum. The T pessary is removed by bringing the bar to the vulva, then rotating one-quarter of a circle, carrying the instrument over towards the natal furrow. This procedure will liberate the anterior bar, and the

* Dr. Scott, of Toronto, Canada, uses my pessaries made of soft rubber, and claims greatest success.

other bar will readily follow. The physician, on his next visit, should carefully go over diagnosis and measurement ; and, when assured that his data are correct, he should replace the instrument, and instruct the patient to withdraw it when it becomes troublesome. The filled T is very readily removed.

When the irritation has subsided, she may introduce it herself again, withdraw when irritability arises, introduce again, and so on. These procedures imitate those of getting accustomed to the use of false teeth. For purposes of cleanliness, it is also advisable to remove and replace the pessary.

Instructions for Replacement may be as follows :

The patient may lie on her back, with shoulders raised upon a pillow. Then, taking the pessary in one hand, enter the loop of the pessary into the vaginal outlet, — concavity of pessary towards left thigh. When the loop has engaged within the vagina, turn concavity of pessary towards the urethra. Bring the hook of pessary forwards as close as possible to the pubis, and, keeping it pressing against the urethral opening, gently slide the instrument backwards till the pessary disappears. In this procedure, the convex edge of the loop slides backwards and upwards on the sacral portion of the vagina, and follows the curve on to the *cul-de-sac.* The danger is of getting the loop into the anterior *cul-de-sac* by slipping it in front of the uterus; hence the great importance of keeping the disengaged part of the pessary close to the vestibulum, or urethral opening. Before using, the instrument should be oiled or soaped. By this coöperation on the part of the patient, I have succeeded in cases which otherwise it would have been impossible to treat, and in others have avoided trouble.

Procedure during Defecation. — 1. Loosen the waist-belt about two inches. This slackens the suspensory rubber cord.

2. Taking hold of the connection of pessary and tubing, gently rotate the hook one-half a circle, and bring it forwards. When this is done, retain grasp of the instrument, and hold it until the bearing down efforts and expulsion of rectal contents have terminated. *In this manner, the uterus is held* in situ naturali *at the very time when the tendency of the forces is directly to aggravate the displacement or cause it to occur, unless prevented by this very procedure. For this reason, it is very important that the patient should remember these directions as they are repeated here.*

When the displacement is slight, the patient may withdraw the pessary altogether, and replace it, as above directed, afterwards.

In case the vaginal irritation is moderate, but troublesome, I have found suppositories of iodoform, grs. ij., gr. x. cocoa-butter, useful before or after the use of the pessary. When the irritation is greater, or is not allayed by these means, the case is unsuited for a pessary; and, if not relieved by procedures to be detailed further on, no pessary should be tolerated or tried.

It is the nature of India-rubber tubing to undergo molecular changes, whereby it becomes decomposed, stretched, and inelastic by contact with the tissues; or, for that matter, without. This point needs watching by the patient. The hook should always be kept half an inch from the perineum. It can only get out of place in two ways: (1) By loosening the belt; and (2) by stretching of the suspensory cord. When these circumstances occur, the patient or physician should refasten the cord to the belt and take up the slack, and then adjust the tension by the belt. If the suspensory cord is a little short at first, the slackness may be taken up by the belt without troubling the connection of

the cord. In giving the patient instructions, it
has been found best not to burden her with too
many directions at once, but to impart them by
degrees as the case progresses. Teach her one
thing at a time; in this way she gains confidence
most rapidly. Women are apt to jump at conclusions at once, and, on the other hand, to become
confused. For these reasons, it is advised that the
physician keep pretty close watch at first of the patient, and afterwards attend at infrequent intervals
as may be necessary. Each case needs special
study. It is not well for the doctor to order a
pessary upon the representation of the patient, and
allow her to apply it herself alone. I have known
this to be done in one instance. The physician in
attendance ordered a retroversion pessary without
any exploration of the vagina: he guessed at the
length. Was it any wonder that the pessary did
not fit, when, on examination, it was found to be a
case of *anteversion?* The patient in this case was
not relieved by accurate treatment, which was de-
layed too long, but became hopelessly insane from
the uterine deviation, combined with other causes.
It is fervently hoped that no one will attempt to
use my pessaries in such an inexact and unscien-

tific manner, as nothing but reproach and disgraced reputation must be expected to follow such unwarrantable usage.

General Treatment. — This local treatment does not relieve the physician from making a special study of the case in relation to general health. This should be conducted on those principles which his judgment approves. No medicine should be given without reason. Indeed, it is safe to trust to the dictates of common-sense, enlightened by study and experience : attention to the trite facts, that life is but a manifestation of force, — vital force ; that life is maintained by the expenditure of force, — dynamic power derived from the sun through food ; that this force is manifested through the combination of carbon and hydrogen with oxygen, *i.e.*, through combustion ; that each individual has a limited amount of vital force ; that the performance of functions reduces the amount of this very force ; that diseased actions and conditions exhaust the stock of forces still more, and that thus version and flexion are drains upon the system's forces, — so that, in order to cure the patient, she must have vital forces increased as much as possible, so as to take advantage of the relief afforded by mechanical means. In this light, the subject of diet becomes one of immense importance. Indeed, it takes precedence of every other. Food is like fuel. You cannot run a locomotive without the production of heat. You may have the noble machine perfect in all its parts, connections, and arrangements ; but, fireless, it is useless for its destined work. So we may relieve the derangements of the uterus, and have the system in a good working condition ; but, unless there is assimilated food-fuel to warm, enliven, and maintain the evolution of the nerve forces, all the perfection of mechanical arrangements amounts to nothing. The writer has regarded this of so much importance in almost every case of these diseases, that he puts his patients upon a diet list, somewhat similar to the following, called the Mixed Diet, — a list from which any patient may select articles of food which have a direct tendency to build up the nervous system. It will be found useful and suggestive. Sometimes it would be better not to undertake any local treatment until the patient has lived for some time upon this diet list, and gained strength enough to work upon.

MIXED DIET.

EAT ANIMAL FOOD.

BEEF STEAK.

Sirloin Steak.

Porter House Steak.

Roast Beef.

Corned Beef.

Cold Pressed Corned Beef.

Smoked and Dried Beef.

Beef Tongues.

TRIPE.

Ox-tail Soup, without Potatoes.

Veal.

Calves' Feet and Heads.

Pork, Fresh, Salt, and Corned.

Pigs' Feet and Heads.

Sausages, properly made.   Ham.

MUTTON.

LAMBS' TONGUES.

Venison.

Turkey.

Game.

Chicken.

Geese.

Pigeons.

SQUABS.

MILK.

BUTTER.

EGGS.

CREAM.

Cheese.

VEGETABLES without or with little starch.

Cabbage.

Tomato.

Celery.

Asparagus.

Onion.

Spinach.

Lettuce.

Dandelion.

Parsley.

Cowslip.

Radish.

Horse Radish.

Cranberry.

Turnip.

Rhubarb.

Squash.

Carrot.

Pickles.

Sour Fruits.

Apple.

Pear.

Melon.

Nuts.

Irish Moss.

FISH, Salt and Fresh.

Fresh and Oregon Salmon.

Cod.

Haddock.

Eels.

Scup.

Perch, &c.

Oysters.

Scallops.

Shrimps, &c.

Halibut.

Trout.

Sword-fish.

Cusk.

Lobsters.

Clams.

Tongues and Sounds.

Wheat Whole.

Wheat Cracked.

Wheat Steamed.

Wheat Crushed.

Wheat Meal baked like Oatmeal.

WHOLE WHEAT ATTRITION FLOUR.

ARLINGTON WHEAT MEAL.

CARR'S GRAHAM FLOUR.

Wheat Bread.   Biscuit.   Cakes.   Crackers.

Doughnuts.   Pies, &c.

Groats.

Oat Meal.

Hulled Oats.

Cracked Oats.

RYE.

Rye Meal.

BARLEY.

Barley Meal.

INDIAN CORN. Meal. Maize.

Hulled Corn.

Hoe Cake.

Indian Pudding.

Hasty Pudding and Milk.

Buckwheat.

Beans, baked, stewed, steamed, or boiled.

Peas, baked, stewed, steamed, or boiled.

AVOID

STARCHES and SUGARS.

COMMON WHITE FLOUR in all and every form, *viz.:*

Bread.   Biscuit.   Cakes, all kinds.   Crackers.

Wafers. Doughnuts. Puddings. Gruels.

RICE, &c.

POTATOES in any shape or variety.

Boiled.

Steamed.

Fried.

Mashed.

Baked.

Roasted.
Irish.
Carter.
Rose.
Lady Finger, &c.
Sweet Potatoes, &c.
SUGARS.
Brown.
White.
Loaf.
Fine.
Coffee Crushed, &c.
Corn Starch.
Maizena.
Arrowroot.
Sago.
Tapioca.
CANDY.

Drinks are food, and may be used according to judg-ment, except in excess, or with sugar.

Air is food, and should be pure.

In all cases the articles of diet should be healthy, and properly prepared. No one should use food that is diseased or infected with parasites. Roasting, broiling, and cooking by steam are preferable methods. Those desiring the greatest benefits should not deviate from the lists. A faithful following of directions insures the best results as a general rule.

In case there is no appetite, quinine, iron, wine, and mild tonics may be administered. In case the food does not digest,— a very common symptom, as the system in the case is too tired to digest properly the food, — lactopeptine, or some of the pepsin preparations, may be given after eating. If there are signs of hepatic trouble, antibilious remedies may be administered; and so on.

Acid Baths. — The medical profession and the public are greatly indebted to Dr. J. H. Salisbury, of Cleveland, Ohio, for the introduction of acid baths into the treatment of chronic diseases. It is not that they were introduced by him into the treatment of *any* disease, but that he pointed out their use in almost all chronic diseases, as a tonic to the skin, an organ which most physicians overlook in the regimen of these cases. The doctor called attention to the fact that the skin is the largest eliminative gland in the body. It is said that if the ducts of the cutaneous sweat-glands were all unravelled and connected together at their ends in a straight line, they would reach a distance of twenty-five miles; that the skin is vicarious in its functions with the internal surfaces and organs; that if there is any weakened surface or organ (whether by diseased condition or functional derangement), then, in case also the skin is subjected to the action of cold, if any " cold " is taken, the effects of that cold will be felt by the weakened surface or organ, be it the mucus membrane of the nasopharyngeal space (as in catarrh), or the mucus membrane of the bronchial tubes in bronchitis, or of the bowels after severe dysentery, or, to bring our subject nearer home, upon an inflamed, irritable, weakened womb. Now, it is exceedingly desirable, in any such chronic cases, to avoid kindling up any return of such diseased action every time that a patient takes cold. It constitutes one of the most serious drawbacks in the history of the case. Dr. Salisbury accomplishes this most successfully by the use of the following prescriptions : —

R   Nitro Muriatic Acid, 2 ounces.
 BATH. One teaspoonful to one pint of water.
    Bathe all over, night and morning.
    Dispense in glass-stopped bottles.

R   Nitric Acid, 4 ounces.
 BATH. One teaspoonful to one pint of water.
    Bathe all over, night and morning.
    Dispense in glass-stopped bottles.

R Aromatic Sulphuric Acid, 2 ounces.
 BATH. One teaspoonful to one pint of water.
    - Bathe all over, night and morning.
    Dispense in glass-stopped bottles.

Where these are employed, — any one may be
used indifferently for the purpose, — the effect is to
give such a tonicity to the skin that it is invigorated
and strengthened so that the patient rarely "takes
cold," and hence gains time for the weakened organ to become strong ; for every time the organ is
vexed or "riled up" by the inflammatory processes,
it is so much the more directly weakened, and rendered more sensitive to outside disturbances.　Another thing, the human skin (humiliating as it is to
confess it) is, in chronic diseases especially, the seat
of the growth of microscopic vegetations, the spores
of which a first-class objective will detect among the
dead *débris* of the epithelial scales, the evaporated
sweat-salts, and some bodies which appear like
cystine.　The acid baths will clear these up and
off the skin, and, affording no nidus for them,
confer a feeling of elasticity and relief to the patient, whose vitality is somewhat sapped by these
parasitic accumulations.

The action of these acid baths is also to check

profuse sweating, — another element of weakness in some cases. They are the most efficient means to prevent "night sweats" in consumption I have ever known. I am not aware that these baths are needed for this purpose so much in uterine disease as in phthisis, but the condition of coldness of the skin, so often found, is very much ameliorated by them.

Their invigorating influence is generally communicated to the internal surfaces, and often they will of themselves be sufficient to start an appetite, promote sleep, and throw a delightful feeling of comfort over the tired and enfeebled patient. These baths may be used with great advantage in many cases. I know of but few remedial agents so positively powerful for good as this recommendation of Dr. Salisbury. No language is adequate to express the high estimation which I have formed of the beneficial effects of these simple remedies, as experienced in several years' use of them in my own practice. I predict for them a universal adoption by the profession, wherever there is a patient with weakened organs in the great cavities of the body, that are disturbed by the patient's taking cold through skin exposure to drafts of air, or involuntary baths of cold water.

The Features of the Retroversion Pessary, to recapitulate, are as follows : —

1. A posterior suspension from a point of support over the stablest part of the body ; namely, over the sacrum, in the median line, through the natal furrow, starting from the belt about the waist.

2. *Single suspension,* so that the deviation of displacement can only be downwards, the natural and desirable way.

3. *Elasticity,* which imitates the normal tone and rebound of the normally-extended and transversely-contracted vagina.

4. A perfect fit around the perineum through the vulva, along the longitudinal axis of the vagina, and into the post-utero vaginal *cul-de-sac,* maintaining the normalities of the long vaginal axis, and allowing and inciting that of the transverse muscular fibres.

5. Unirritating material.

6. Selection and application by mensuration.

7. A contrivance to hold the uterus *in situ* during defecation.

8. It is within the patient's control.

Why use an External Support ?—It is clear that an intro-vaginal pessary, extending the vagina in

its long axis, must have some point of support from
within the vagina.   Its base must rest on the tissues
near the vaginal outlet.   Generally, the pubis re-
ceives the pressure, there being a rentrant angle to
receive the urethra.   It also rests upon the middle
of the labia majora, inside.   If the pessary is so
small as not to distend the vagina laterally, it
would not take a bearing, and would soon be
ejected by the natural expulsive efforts of the
parts.   There must, then, necessarily be a *lateral
distension of the vagina.   But this lateral distension
is a prominent pathological element in the complaint.*
To cure, we must relieve it, and avoid it, and induce just the opposite condition.   The line of support should be in the central axis, but the pubis
and the perineum are outside this line, — they are
peripheral ; hence should, for this additional reason,
not be employed for this purpose.   The writer may
be mistaken, but it appears theoretically impossible
to have an extension of the vagina longitudinally,
and a non-distension of the transverse diameter,
without an *external support.*   He is aware that this
is the opposite of the commonly-received opinion
held and acted upon by the profession hitherto.
The great majority of pessaries now in use are

all intra-vaginal. They present a great variety of mechanical shapes, so diverse and grotesque and anomalous as to be a surprise and wonder to any one who has patiently considered the physical conditions of the normal vagina and uterus. It has long seemed to the writer that many of the common pessaries were invented in fits of despair; the inventor seizing upon any thing that he might by chance discover. May we not be allowed to ask whether this very ignoring of all the elements of the physical problem of uterine support may not be in part the cause of the opprobrium that rests upon all pessaries? Why should we hear them condemned vigorously as useless, and worse than useless? Do we hear such complaints about other mechanical appliances to the human body, such as coats, shoes, gloves, &c.? No; because the artists who manufacture these articles pay some attention to the laws of fitting and moulding, which we have shown are violated whenever the attempt is made to support a retroverted womb by not restoring the metamorphosed vagina, the same having been altered diametrically as much as possible, — the long diameter having been changed into the short, the short into the long, — and the whole concern having become a sort of a hollow, wrinkled bag filled with the womb, and air or gas, when it should be a tube with antero-posterior walls in contact, and the neck only of the womb projecting a little into it at the top! Why, then, adhere to a practice that involves a physiological and mechanical error? We well know that a pessary that is inside, out of the way, involving no care of the patient, is a very convenient and satisfying instrument, because it is very much less trouble to care for it and manage it. We record our admiration of the ingenuity and simplicity of the pessaries devised by our late, honored, beloved, and Christian preceptor, Prof. H. L. Hodge, M.D., of the University of Pennsylvania. We have seen cases of great relief from their use; but the very candor, the love of truth, and the appreciation of it, which he inculcated upon the writer, demands of him fearlessly yet respectfully to point out what he feels to be errors of judgment, errors of mechanics, and errors of application in regard to intra-uterine pessaries. It is gratifying to know that these intra-uterine pessaries have afforded some relief; and if they — constructed on false principles — have done so well, what greater things can we not expect to be accom-

plished with pessaries that avoid their errors? I
am far from saying that my own contribution is the
best that can be devised on these principles; I am,
however, happy in saying that I believe their principles are correct if their application is not.

Why a single support?— Because there is less
chance for displacement. Simple things are always
preferable if they will do the work. It simplifies
the pessary very much to have *one support only;*
takes less material, has less detail and less attention. If there were two points of support, — one
in front and the other behind, — there would also
be a want of correspondence, during the body
movements, between the motion of the abdomen
and that of the sacrum. It would be troublesome
to adjust the bearings equally, to get the tension
right on both; while the single support is very
readily adjustable, having nothing but a single element to arrange. If one support answers, why
use two or four?

Why posterior?— The sacral region behind is
the broadest, smoothest, firmest, and most stable
bony point in the external surface of the human
body. The sacrum is the keystone of the arch
which holds up the vertebral column, — the basis

of the whole trunk. Compared with the abdomen as a point of support, it is preferable, because the abdomen is subject to a constant motion in respiration, and to a varying size, because of distension from flatulency, food, fat, and foreign bodies, and of flattening from the evacuation of the intestinal contents. The urinary bladder lies underneath the point of abdominal support, varies in size at all times, and often is sensitive to pressure. These conditions do not affect the sacral surface of the pelvis. The pessary surrounds the sacrum in front and behind. When the sacrum moves, the pelvis moves, and the pessary with it. There is no relative displacement, then, from causes that vary the position of the sacrum, which affords a firm, naturally non-sensitive point of support.

Another reason is to interfere as little as possible with the excretory discharges. The function of defecation is generally very much less frequent than that of micturition. The posterior support does not interfere with micturition at all ; and, during defecation, the arrangement is such as to remove the perineal hook out of the way, and also afford a sure means of support during the bearing down efforts of expulsion, which, of all agencies, make a retroversion worse, besides hindering the passage of the fæces by being pushed and packed against the rectum.

Another thing, — the posterior direction is much the shorter distance to the waist belt. The vaginal outlet is nearer to the back than the front. It is a common principle in mechanics to get the power and weight as close and direct as possible ; so here, the lines of elevation and support meet at an acute angle or curve, thus approximating the force and the body to be moved as closely as possible. If suspended anteriorly, the curve would be broad, the angle obtuse, the direction of the forces not so controllable because so indirect and distant, and often interfering with micturition.

Why an elastic support? — There is a certain amount of natural uterine mobility during coughing, lifting, and even breathing. A careful observation, during a speculum examination, will show a play of the uterus corresponding with the movements of the diaphragm. The natural spring-like elasticity of the parts on which the uterus is suspended bring it back to place. The sphincters of the vaginal column, acting together, and resting on the perineum, act, as was said before, like the coils

of a spiral spring; its bound and rebound, acting
with the diaphragm, resisting the bearing down efforts in lifting, straining, or running. And it is easy
from this to see why these efforts are productive of
so much trouble, when the uterus has become displaced down into the middle of its expanded and
"squashed" spiral spring, — a dilated, weakened, and
metamorphosed vagina, — and how helplessly pros-
trate it must become in time under such physical
conditions. When the springs of upholstery work
are dilated, expanded, and flattened, the seat becomes inelastic and uncomfortable; indeed, worse
than with no springs. We can discard these, and
replace them; but the problem in retroversion is
to restore the vaginal spring during its very work,
— not to put in a new one. But the muscular fibres
(sphincters) have been dilated so long a time as to
be perfectly useless for the purposes of elasticity,
recoil, and rebound; hence, we must supply the
elasticity from without. In order to imitate it in
the artificial support, India-rubber tubing was selected as approaching more closely than any other
artificial substance the tone and rebound of healthy
muscular and connective fibrous tissues. It might
be possible to construct the pessary itself of a

covered spiral spring, but it would not do away
with the elastic suspension ; as a metallic or hard
substance in the natal furrow would be hurtful,
by bringing the skin between two hard bodies, the
sacrum and the metal or hard rubber. The present
arrangement of the spring-suspension outside has
proved admirably satisfactory. Perhaps the addition of a spring within the vagina may be desirable
in some cases, as it is perfectly consistent with the
principles of restoration. The suggestion occurs
to me as I write, and is noted down as a suggestion for future use and consideration. The India-
rubber tubing is unirritating generally to the natal
furrow, and allows of a healthful amount of play
within certain limits. The elasticity of the tubing
varies with the quality. It should be examined at
intervals ; for, if allowed to stretch too far, displacement will occur. These defects may be prevented by proper care in taking up the slack.
The lower part of the cord wears out first. It
may be reversed end for end with advantage and
economy. Sometimes the part that grasps the end
of the pessary becomes expanded and loose from
use. This end may be cut off with scissors, and
the cord re-attached to the pessary at a vital part.

The liability to displacement of the retroverted uterus after adjustment may be estimated as follows : —

Downward. — The suspensory cord obviates this when it is of proper integrity of substance, and sufficiently taut. It is the one most liable to occur, as all the forces of abnormal and normal conditions tend downwards.

Upward. — The perineal hook prevents this. It is seen in cases of long standing, where it is not possible to restore the normal length of the vagina at once. The pessary will be found to fit very well for a while, and then to hurt or incommode. The vaginal sound will soon detect this difficulty, and indicate the proper sized pessary ; hence, the desirability of examining the patient occasionally.

Forward. — This can only occur when the pessary is too long ; and, if the uterus is quite mobile, the retroversion will tilt over into an anteversion. This happens sometimes, not often ; and is generally detected by *objective*, not *subjective*, symptoms. It is more liable to occur in the cases of largely dilated vaginæ.

Backward. — This is prevented by the pessary and the sacrum. But sometimes this will partially occur, and the uterus bend backward (double over), the pessary turning towards the sacrum. In these cases, the uterus is relaxed and flabby and toneless, like a freshly made clay-vessel in the hands of the potter. These are troublesome cases. Unless this flabby condition can be relieved by improving the tone of the system, it will be necessary to use a stem pessary, to be described farther on. I have noticed that this condition varies from time to time. On some days it will disappear, to reappear some days later. It is not a common complication, but occurs often enough to be looked after and recognized.

Laterally. — This is prevented by the loop around the cervix, and engaged in the *cul-de-sac.* Indeed, when the uterus and vagina are in the normal site, the broad ligaments and the utero-vaginal attachment of themselves strongly tend to keep the womb in place. The artificial support is antagonized by these attachments. To make it more clear, pardon a homely illustration. If a model uterus were made, large enough to fit into a coat sleeve, and attached to it, imitating the natural vaginal attachment, — should the model be held or pushed onward until the sleeve became tense, it is evident that the resistance of the sleeve would erect the model into its natural position; just so with the womb. These lateral displacements do not often occur, and can only be detected by the introduction of the uterine sound. The vagina must be very largely distended; so that, unless the pessary is a very broad T, it pushes to one side in a transverse fold, and thus causes the womb to deviate also. However, this deviation is practically of minor importance as to disturbance, recognition, or frequency; and is alluded to in order to make the description complete.

Remarks in Relation to the various Classes of Retroversion, and in Regard to Treatment. — They may be divided into, — 1. Ordinary; 2. Extraordinary; 3. Impossible.

1. *Ordinary.* — In these cases the uterus is mobile, normally long, not sensitive, engorged, instable, flabby, and of normal consistence and resistance. There is no abnormal condition, — perimetritic, inter-metritic, or intra-metritic, — except what is directly connected with the retroversion. The procedure in these cases is direct, simple, and immediate, and embraces the replacement of the uterus with the uterine sound.

(*a*) The measurement of the distance from the posterior fourchette, along the concave surface of the posterior vaginal wall, to its termination in the *cul-de-sac* behind the uterus.

(*b*) The selection of a pessary one-half inch longer than this measurement.

(*c*) The introduction of the pessary while the uterine sound is holding the uterus as nearly as possible in natural place.

(*d*) The removal of the sound after the pessary is placed.

(*e*) The holding by the physician of the suspensory cord behind, while the patient rises and stands up until the belt is adjusted around the waist.

(*f*) The examination of the suspensory cord after the belt is fastened, to see if the tension is right. Often the cord is too long, — left so purposely ; as it is easy enough to shorten it, but not to lengthen it. This tension arrangement needs the closest attention of the physician, as it is the vital feature of the treatment. The pessary may be fitted accurately to vagina and perineum, but if the tension of suspension is at fault, harm results. If it is too tight, the pessary will injure the perineum or the *cul-de-sac*, or antevert

the womb. If too loose, the uterus will settle
down, the vagina become wrinkled, and the pessary
will be apt to slip out of the vagina, and dangle
uselessly between and behind the thighs.

(g) The patient should be requested to walk
about the room, sit carefully down on a chair or
cushion, to see if the instrument hurts her. If it
does hurt much (and one must not confound the
pain which sometimes follows the introduction and
use in elevation of the uterine sound with that
produced by the pessary, which does not have that
peculiar exquisite and sickish feeling which results
from touching the fundus of the healthy womb),
then wait a little while, cause the patient to lie
quietly in bed, and, if the pain does not then subside, have her withdraw it; and on a subsequent
day the physician should go over his data once
more to ascertain if there is any mistake. But
in this class of cases, it is very unusual to have
any trouble; the pessary fits equably, and confers a
feeling of relief, which renders the patient very
grateful, and enables her to pursue her avocations
and walk about without trouble, because her womb
is not jogged down by every step she takes. If
no other gain were made but this of the locomo-

tion, the pessary would justify its invention and application.

(*h*) The patient should be instructed according to directions already given; pages 29 and 37.

(*i*) The physician should call, in a short time, to see how things are, and impart further directions.

If the anteversion is of a few weeks or months standing, permanent relief may be anticipated by wearing the pessary a similarly short time, and then discarding it. But if it is an old case, relief must be ensured by a long wearing of the instrument, until the lengthened round ligaments regain their normal length by absorption, and by those nutritive processes which involve molecular changes of long duration and slowness.

2. *Extraordinary Retroversion.* — These cases are of a long-standing; uterus enlarged, sensitive, sore, ulcerated perhaps. The vagina sympathizes in irritability, and the digital exploration is apt to produce agony (vaginismus). In these cases, make haste slowly. By no short and easy process, — like Elijah's healing of Naaman, the leper, — can we hope to effect a cure here; but the disease must be, to use a common word, "courted." The gen-

eral health must be improved by means already
suggested ; page 40.

~ Vaginal or rectal suppositories of iodoform, of
belladonna and opium, of morphia, &c., may be
employed to allay irritation. If they do not succeed, — no matter if the general health does not
seem to warrant it, — put the patient to bed, and
use local depletion ; unload the congested capillaries
which have been so long unnaturally distended
that they seem in a measure to have lost their
normal power of contracting, and thus of rightly
propelling along the stream of blood which consequently stands therein, — allowing of osmotic
extravasations of the serum, migration of white
blood corpuscles through the fenestræ of the capillaries, and thus producing what is called congestion, sub-acute inflammation, irritability, &c.
Now it seems — and this principle is just as true
of some catarrhal conditions of other parts of the
body — as if the most sensible procedure in such
cases would be to draw off the contents of those
capillaries, in which case they would collapse and
contract of themselves by having nothing within
to resist ; in the same manner as the tube of the
vagina, distorted and distended by the retroverted

uterus, has a chance normally to contract on itself
when the uterus is elevated to where it belongs.
It is wonderful how rapidly this remedial process acts sometimes. I have seen a uterus that
measured four inches inside reduced to three
inches, by the application of three leeches, in
less than two days' time! It appeared not only
that the organ was drained of superfluous blood,
but that also the relieved capillary interstitial vessels went immediately to work (as it were) to absorb the products of effusion, or hypertrophy, by
reinstitution of the normal processes of nutrition,
whereby the normal balance of supply and waste
of tissue, and the preservation of the bodily organs
at their usual size, are maintained. Be this as
it may, the writer knows of no procedure equal
to local depletion in the extraordinary cases of
version.

Once I had a patient who had a verted uterus.
She was afflicted with a congenital stricture (hereditary) of the æsophagus, so that she could eat no
solid food whatever. Consequently, she looked to
a stranger like a person in the last stages of consumption, — pale, anxious, thin (bones staring one
in the face), weak, and feeble. Her uterus was

badly ulcerated about the os, and discharged large quantities of leucorrhœal secretion, which, under the circumstances, were very weakening. For months, I treated her without depletion, as it appeared so thoroughly contra-indicated. Still, the uterus was sensitive and enlarged, the monthly flow was very copious and exhausting, and my treatment did not amount to much. After a while, I was induced to reason thus: unless those capillaries of the uterine tissue are relieved of their contents, the menstrual normal congestion, superadded to the abnormal one, will continue to cause the profuse menorrhagia, keep up the ulcerations and leucorrhea, and my patient will be reduced as much by these processes, without advantage, as she would be by local depletion. Furthermore, if the menstrual epoch finds the uterus free from congestion, there will consequently be less menstrual hemorrhage ; and hence less exhaustion, less driving of the wedge of the' disease tighter. It was somewhat difficult to persuade the patient of the propriety of the procedure, but finally she consented ; and the application of two leeches to the os caused a profuse hemorrhage, but not comparable with the menorrhagia.

The uterus in this case, also, was reduced one inch in length of cavity during forty-eight hours. The leucorrhœa diminished, the sensitiveness disappeared; and for a long time subsequently she wore her pessary without trouble. She lay in bed a few days after the depletion, feeling somewhat weak, but very much less than at the meno-flow. So that, from this very discouraging-looking case, I have been greatly encouraged to carry out this principle of relief by local depletion; and have since tested it in many other less forbidding cases. I should add that this patient, being the overworked mother of a family of six, would get exposed, take cold, and suffer a renewal of the congestion of the womb, which would be relieved at intervals by the leeches.

Methods of depletion of congested uterus by abstraction of blood contents: —

Scarification. — This may be produced by pricking the uterine neck with the point of a bistoury, or little spear-head. The blood will flow more freely, if a dry cup is applied beforehand. The spear-head may be deeply thrust into the substance of the womb, and the hemorrhage encouraged by vaginal injections of warm water.

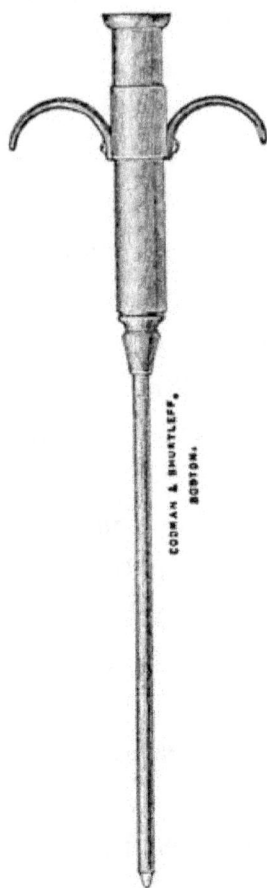

Fig. 9. — Cutter's Uterine Scarificator.

Fig. 9 is a uterine scarificator, adapted to an automatic plugger made by Codman & Shurtleff. The common intrauterine scarificators of Pinkham or Miller may be used inside the uterus at the will of the operator. (Figs. 9 *a*, 9 *b*, and 9 *c*.)

Leeches. — I like them best, and employ them when they can be obtained. Their bites cause a long and slow flow of blood, — just what is needed. The objections to these are : —

(*a*) The aversion of the patient. This may be met by the promise of the physician to give his personal attention and care to their application.

(*b*) Their not taking hold. This is the most serious objection. A lively leech will often, the moment it gets within the vagina, find it has an urgent business call outside, and make a lively exit immediately. If prevented, sometimes it will fasten on to the tube of the speculum, in the vain effort to get blood out of glass. Sometimes it will curl, and take a *siesta*. Sometimes it crawls in by the end of the speculum, and out between it and the vagina; and so on.

(*c*) It has seemed to me that leeches are very sensitive to odors. They do not like the smell of soap, scented or not, of the vaginal secretion (the vagina should be well syringed before applying a leech), of lard, glycerine, or any common lubricant. Therefore, the writer uses only clean, pure, warm water as a lubricant. The vulva is well moistened with water; and also the speculum, which is then very slowly introduced. It is harder

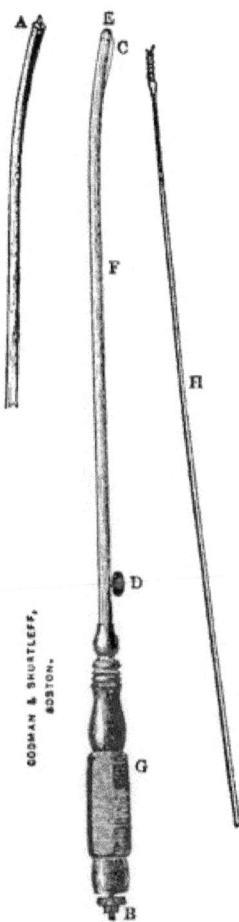

Fig. 9 *a*. — Pinkham's Scarificator.

for the patient to bear; but a steady, gentle hand will avoid much hurting. The speculum I have lately used is a glass one, improvised from an atom-

CODMAN & SHURTLEFF, BOSTON.

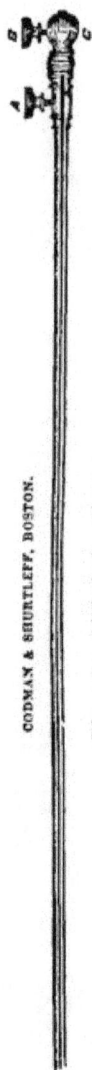

Fig. 9 *b*. — Miller's Scarificator.

Fig. 9 *c*. — Miller's Scarificator.   Uterine end; full size.

izer face shield. The large, expanded shield makes
it easy for the patient to hold the speculum in place,
as the bearing surface is so broad; and the contracted extremity makes it easy of introduction
(Fig. 10).

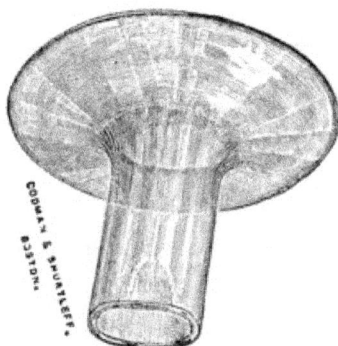

Fig. 10. — Speculum.

The patient may lie upon her left side, knees
drawn up, against a window, curtained upon its
lower half; or, at night, the light may be thrown
into the speculum by a reflecting mirror on the
forehead (Fig. 11).

(*d*) The leeches, placed in the speculum or in a
leech-tube, may crawl into the womb. This is a
well-founded fear, as accounts have recently reached
us of a leech actually living in a man's larynx
for several months! I have never known of but
one instance of a leech getting inside the womb.

In this case, it took hold just above the internal
os, and filled itself; then let go, and came off

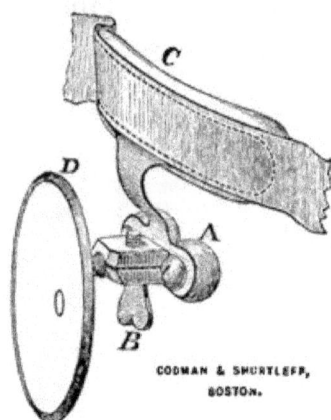

Fig. 11. — Head Mirror.

and out of the womb of its own accord. Should
an accident occur of the leech going into the
womb, a solution of common salt and water would
immediately cause its withdrawal; and it may be
that the salt in the vaginal secretions prevents
their taking hold at times.

(*e*) It is necessary for the physician to know
how many leeches he introduces, — keep count of
them, watch till they come off, then replace them
in their box, or give them to an attendant.

(*f*) Generally they can be relied on but for one
use. Three is a moderate number.

These are six of the greatest objections against
the use of leeches ; but, in my own mind, the
prompt and positive results they induce are a sufficient reward for all the pains they cause.   For
example ; lately, a lady with a verted womb, tender
and sore, with headache and gastric distress severe, had two leeches applied to her uterus.
The result was an entire loss of the headache and
gastric trouble, — which latter had existed for eight
years !   And what other .means will reduce a hypertrophied womb so speedily ?   I know of none ;
and, knowing none, I am not disposed to relinquish
a tried and successful means, even if it is hampered
with six objections.

These methods presuppose an intra-vaginal application of the leeches.   They may be applied outside
by the patient herself, in some peculiar circumstances.   The sites of application may be over
the bladder; over the ovaries (just inside the
anterior-superior spinous processes of the ischia);
or inside the thighs, near the vulva.   The outside
application takes double the number of leeches.
They are very easily applied by inverting the
opened box, in which they are kept, over the
site of application.   In cases of extreme vagin-

ismus, and when the patient cannot be waited upon by the physician, this outside use of leeches may be resorted to.

Having, then, removed the irritable, abnormal condition of the womb by the local depletion, a few suppositories may be employed, and then the physical exploration of the vaginal neighborhood and contents may be conducted ; and if the diagnosis by exclusion reveals no difficulty but the retroversion, the case may be proceeded with as an *ordinary one*, only watching more closely and carefully. The patient may be kept in bed while testing the pessary. This is useful, because it insures an even temperature of the skin, avoids exposure to draughts, and husbands the vital forces for the work in hand, — which I am accustomed to inform my patients is as much of a *business* as doing housework, or running a sewing-society. The patient must feel that she is *really doing a great deal by keeping still*, although she may not see it. At any rate, the business of *worrying* is *more* exhaustive than any other, and pays the least dividend. Do not worry !

(3). *Impossible cases* of Retroversion are those that are unmanageable. It is lying to say there are none such. These are they that have long

existed under aggravated circumstances, and that
have caused incurable insanity, or such a hyperæsthetic condition of things that the patient's life
is one of continuous agony. In some of these
cases, the derangements of the nervous system
would continue, even if the retroversion — the
primal cause — were entirely removed ; as the system has become so habituated to it, that the abnormal conditions have become a second nature. This
third class includes a retroversion caused by or
complicated with adhesions or tumors, malignant
growths, pelvic abscess, prolapsed ovary, fibroids,
sarcoma, carcinoma, aneurism, hematocele, diseased condition of the bowels, omentum or peri-
toneum ; indeed, any of those irremovable physical
conditions, either extra-mural, inter-mural, or intramural, which are sometimes found associated with
a verted womb. In these cases, a pessary cannot be tolerated ; ought not to be tried ; is perfectly inadmissible. The only range of conditions
is where treatment can abolish the hyperæsthesia
of the uterus so that it will bear handling, and
when the womb can be elevated into its place
by mechanical means that are unirritating as long
as they are employed. Professional skill may

diminish the number of the impossible class, —
it cannot treat the whole of them ; and this fact
may be admitted to the patient at the outset, so as
not to raise expectation unduly.　Some unusual
procedures may be resorted to in this class, among
which we mention, —

Elevating the foot of the bed upon which the
patient lies while she is trying the supporter.　In
this position, the contents of the abdomen gravitate
towards the diaphragm, remove pressure from the
womb, and tend to draw the uterus upwards.　If one
wishes a more perfect and easily managed physical
contrivance, that of the writer may be employed.

Fig. 12. — Cutter's Invalid Chair.

Although this cut does not show this position, the chair can be straightened out like a bed, and the head depressed until it may almost touch the floor, thus giving a single inclined plane whose inclination to the horizon is easily, gradually, safely, and certainly controlled by the endless screw at the side. It may also be arranged into a double inclined plane and a triple inclined plane. Having the motion of the great joints of the body in vertical planes, the chair can be adapted to almost any position of resting on the back.

There may be means of allaying these hyperæsthetic conditions not known to the writer, of course. He has not employed, for instance, the anæsthetic influence of carbonic acid gas, as used by Simpson. It is possible that exposure of the irritable vagina and uterus to a bath of this gas may so paralyze or anæsthetize them as to render them capable of being handled without pain ; if so, it would be very easy to generate and apply the gas before using the sound.

History of the Origin of the Retroversion Pessary. It may, perhaps, be interesting to the reader to learn how this pessary came to be invented. It originated with the case of a mother of a large family, middle-aged, doing her own work, — the hours of labor being constantly and continuously, year in and year out (when possible), from five in the morning to eleven at night, — a busy, occupied woman, of great nerve-power, endurance, and vigor. She had a retroverted uterus, which my venerated father treated long before the writer came upon the stage. The symptoms, as remembered, were severe præcordial pain, as if a knife were being stabbed into the heart, a very easily excited nervous system, pain in bowels ; hurt her to lift weights or ascend stairs. Digestion was not impaired. Still she kept on at work, though bearing about trouble enough to wear out an ordinary woman. Following my father's example, after his death, I rang the changes on intra-vaginal pessaries. Still the difficulty was unrelieved, and I had about given up the case in despair, when the thought was given to me, — by some good angel, we may hope, — that the treatment I had subjected this woman to was a failure, because it was founded upon incorrect mechanical principles, and was contrary to the rules and practice of common-sense ; as no human-body appliances in general use were ever applied without some simple but accurate system

of measurement being resorted to in their making
and use : and here I had been trying to hold a delicate but heavy organ in a sensitive cavity of the
body, out of sight, by physical means which were
adopted upon recommendation of my teachers, and
upon an explanation which I never could understand, although I cannot be said to be destitute of
a "natural turn" to mechanical principles ! I was
led to ask, Are there not some principles upon
which a proper treatment can be based ? It seemed
to the writer that the treatment should embrace an
accurate fit ; that this could be obtained satisfac-
torily only by measuring accurately the surfaces to
be extended ; that as men measure people for
hats, coats, and shoes, for instance, — appliances
which are adapted to the outside of the body, and
directly under the eye, — why should a perfect fit be
thought possible without mensuration in the case of
a body-surface which is internal, delicate, and out
of sight ? Having, then, settled about the measurement, I was led to suspend the pessary from behind,
by a single suspension, for reasons already mentioned. The size of the perineal hook was at first
very small ; and here has been encountered the
chief trouble of the pessary, as might be expected.

At present, the large curves are found generally
most comfortable. I may be allowed to say that my
patient wore her supporter for several years with relief ; in fact, it was the *only relief* she could obtain.
If it were left off, the symptoms would return in an
unendurable form, and the uterus would vert back
again. She passed the menopause, and the uterus
remained very well in place when the supporter
was withdrawn. Still — hard at work as ever —
she finds that the severe præcordial pain will occasionally return. Lately, however, the pessary,
which had served so long, did not fit. On mensuration, it was found that the vagina had actually
shortened half an inch! This is the only instance
of vaginal shortening I ever met with. Upon sup-
plying a pessary half an inch shorter, a comfortable fit was immediately insured. It is probable
that the organ had suffered from atrophy, as the
menstrual function had ceased for some time.
When this patient informs me that she would
have been dead or crazy, but for this invention, —
and as I have some reason to believe her statement
contains more truth than poetry, — I am thankful
that I have been instrumental in her relief, and do
not feel unrewarded for my long labor and pains.

For the description of the *subjective* symptoms of retroversion, the reader will please look farther on; and we will close this division of our subject by simply describing the patient's feelings in a case of Retroversion of long-standing, as a sort of illustration of the disturbance a retroverted uterus causes.

The patient was a married woman, about thirty years of age; had two children within five years. Prominent symptoms: For many years, — even before marriage, — headache so severe as not to be relieved except by taking off the skin of the neck with mustard-poultices; this generally occurred a week before menstruation, which was always difficult from the first. Pain in the chest, upper part, front and back; with a sense of suffocation and a feeling of stricture, as if a handkerchief were wound around the neck. Severe pain almost constantly, seated over the heart; sometimes even angina pectoris. Numbness of limbs — left arm particularly — for half a day at a time. Digestion very weak. Always pale and thin, and anæmic. Very profuse menstruation, associated with excessive pain. The sufferings were at times intolerable and beyond description. Was very

much better after her first child was born. The
back is the weak place, and she has suffered here
very much always. It is a dead, heavy pain. No
sense of stricture around the waist. Since last
winter (1874–5), she has had pain over the left
ovary,— a severe, agonizing pain, making her excessively nervous. *No leucorrhœa. No bearing-
down pains in bowels. No trouble in walking or
going upstairs, except from exhaustion and weakness. No trouble in defecation.*

After the womb was restored by a uterine sound
to its normal situation, the thoracic symptoms of
pain and suffocation disappeared, although the
uterus immediately went back into its old site, like
an outstretched spiral spring suddenly let go.
From the negative symptoms, the patient supposed she had no womb trouble; and no one
hinted the existence of the retroversion until this
time, although she had been examined by a
physician whose name is a tower of strength
and fame, and who is thoroughly worthy and un-
usually intelligent. But such is the truth, such
the cause of suffering overlooked. It is possible that an examination may have been de-
clined, on the ground that the discovery might

involve an amount of attention not at command,
and of great trouble, with no corresponding benefit. It is charitable to believe that there was
some reason for withholding a thorough examination other than ignorance.

RETROFLEXION.

Definition. — In this displacement, the uterus is bent backwards upon its longitudinal axis, the cervical portion being in its natural position. There is a varying degree of prolapsus.

Diagnosis. — A digital examination of the vagina finds a dome of continuous hardness extending backwards from the os. The rounded fundus is felt towards the sacrum. Somewhere between the fundus and os is felt the angle of flexion, — a sort of rentrant angle, which varies from completeness, — the uterus being doubled on itself, so that fundus and os are side by side, — to a very slight inclination. The posterior utero-vaginal *cul-de-sac* being absent, in some cases the flexed fundus is undistinguishable from a tumor, — only then the uterus is straightened and elevated into place by the uterine sound. The finger easily traces the uterine direction by following along the posterior surface, and reveals the amount of bending deviation better than any thing else. The point of flexion is not constant. It may exist anywhere, from a point half an

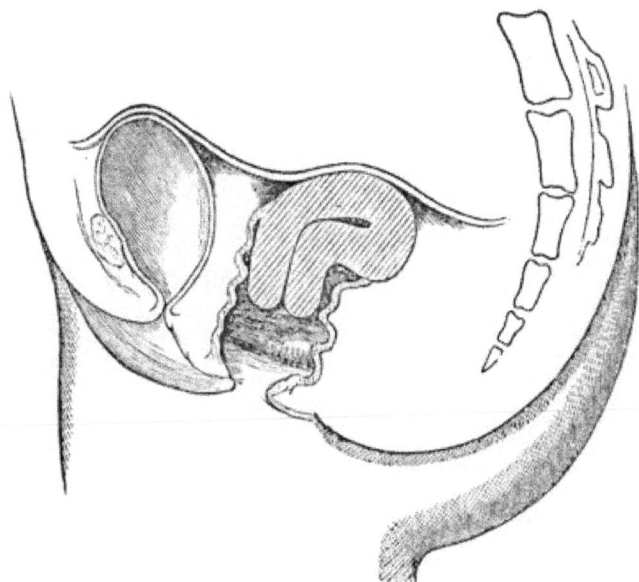

Fig. 13. — Retroflexion.

inch from the os to three-quarters of an inch from the fundus. The diagnosis should be confirmed by the uterine sound. It should be clear and distinct ; as, if there is a tumor or pelvic abscess or malignant disease, no interference to restore the uterus should be made. Unless the sound is bent to a considerable angle at its point, and the handle is carried out far in front between the thighs, it will be found a difficult matter to introduce it into the uterus and straighten out the flexion. When the sound is once entered, it may be withdrawn and the bend lessened, and then re-introduced, until, repeating this process, the uterus is made normal in axis direction. The degree of flexion determines the amount of perimetritic disease, adhesion, neoplasms, heterologous growths, inflammation, and its results, &c. And this information is afforded by the attempt to introduce the sound, and elevate the uterus to its normal site. In this procedure, a fallacy may be produced by the spring-like resistance of the womb itself to being straightened. It is almost like India-rubber when stretched out.

Indications of treatment : —

(1) To remove the flexion.

(2) To restore the uterus to its place.

(3) To maintain it there.

The first two indications are met by the use of the uterine sound. The third, by taking a measurement of the posterior vaginal wall with the vaginal sound, — selecting a loop or T pessary half an inch longer than this measurement, and applying it after the manner described under Retroversion. When

the uterus is elastic like India-rubber, it happens,
that, after the flexion is reduced and the sound
withdrawn, the organ becomes almost immediately
retroverted as before ; and this so after many trials !
In these cases, the most evident and immediate
means of prevention would be to keep the sound
in the womb all the time. This being out of the
question, the writer has adapted a stem to the T
pessary, and introduces it in place of the sound.
When this is done, it mechanically
straightens and acts as a splint;
unless the uterus is much enlarged,
this stem will hold it *in situ naturali.*
If the uterine cavity is much greater
than usual, the stem will push forward and be in place while the
anterior uterine wall is dragged forward, and the posterior wall is at
its old place. In this, the reduction is
more apparent than real, and is mis-

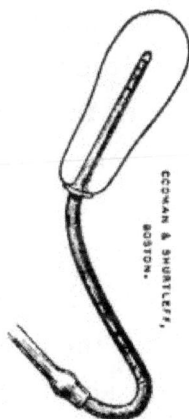

Fig. 14.
Cutter's Stem.

chievous and dangerous. This danger may be
avoided by resorting to measures of depletion
already described to reduce the organ to its normal
size, consistency, and tone. *A cardinal rule* for the
preparation for any of the set of pessaries *is never*

to put a pessary, if possible, in contact with a sensitive womb, outside or inside.

·As is seen in the cut, this stem pessary differs from the ordinary intra-vaginal stem pessary by being attached to, and continuous with, the T pessary; only that the cylinder is shorter, and the bar replaced by the disc and stem.

Vaginometry for a stem pessary is accomplished by the uterine sound. When the retroflexed uterus is straightened and replaced, the length and diameter of the uterine cavity and outlet is noted.

Next, the distance from the os to the perineum is also noted. Some idea of the perineal thickness should be gained by palpation. A pessary should be selected which has a stem of two and a quarter inches in length, and a thickness corresponding to that of the cavity. The cylinder from the disc to the inside of the curve of the hook should be one half inch longer than the distance already noted from the perineal edge to the os. The curve of the hook should correspond with and be a little larger than the curve of the perineum.

Application. — Immediately on withdrawing the uterine sound, the stem should be passed up over the forefinger to the os, and introduced by gently pushing upwards. This reads more easily than the performance. Of all hard things, the introduction of a stem is a marked one. The difficulty lies in having the disk of the stem strike upon the perineum, and catch painfully there. With a finger and a disk the vaginal outlet is pretty much occupied, so that it is necessary to engage the pivot of the stem in the os uteri, push it in until the disk strikes the perineum, then withdraw the finger, bring the perineum outside the disk, and push the stem in as far and as gently as possible. When the disk has entered the vagina, the finger then may be passed in beyond the disk, and the amount of stem-entrance observed. But it is a rather painful process. Perhaps a stem might be arranged so that the disk could be put on after the stem is already in, — this I should deem a desirable improvement. In this case the disk would go in first, then the stem through the disk into the womb ; or the disk might be placed on after the stem has been introduced into the womb, by being slotted largely, and then rotated about its axis and held in place by a catch. Sometimes it is impossible to introduce the uterine sound into a flexed womb. This is due partly to a very flabby condition of the uterine walls, so that when the sound is passed up to the angle of flexion, it stretches the concave wall upwards and pockets itself there. If this is not overcome by alteration of the angle of the bent point of the uterine sound, it is best not to try too long, as the mechanical irritation may provoke inflammation, and stop all attempts at restitution. I have found that by waiting a few days the uterus may take on more tone, and then the sound may be

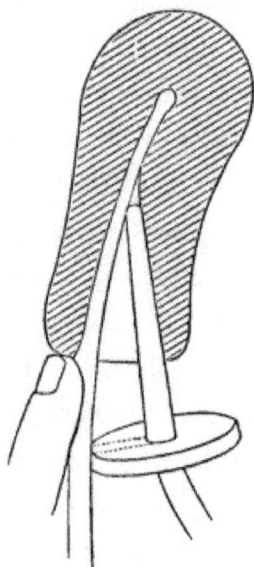

passed in ; because the concave uterine wall resists the pocketing process, and, as the sound slips over it, the convex uterine wall is drawn up and the flexion straightened. In these cases, and in those perverse ones that bend back immediately on the removal of the sound, the stem will not enter, of course, when the uterine sound is withdrawn, because the stem is more difficult of introduction. (Fig. 15).

Fig. 15. — Application Stem
Pessary.

The plan is, then, leaving the uterine sound in the womb, to pass the stem in by its side as far as it will go without injury. Then withdraw the sound; and, as this is done, advance the stem forwards. Generally, this with me has not failed of success. When the stem lacks less than half an inch of thorough introduction, if allowed to remain, it will be found to advance into place of itself by the next day, as the tension of the suspensory cord is constantly in this direction. When these procedures fail, a slot (*vide* dotted lines in Fig. 15) should be cut into the disk of the pessary in front, so that the stem and the sound come in contact with each other, and avoid the stretching which occurs in the use of the unslotted disk. This procedure has rarely failed with me. When it has done so, I have been obliged to acknowledge myself vanquished, and the case incurable, so far as the writer is concerned; or, have sought to build up the general health, and then make another trial.

Since the above was written, I have devised the pessary (Fig. 16) which practical ex-

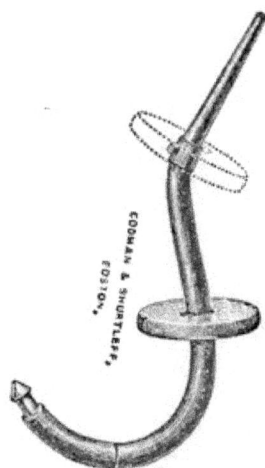

Fig. 16. — Cutter Stem; Movable Disk.

perience has shown to be a great improvement in
the facility of introduction.   The disk is made movable upon the cylinder.   At the base of the stem
is a projection.   The disk is provided with a recess
to receive this projection, cut all around the circum-
ference of its central opening.   A notch in the edge
of the disk indicates the point where the projection
may slip into the recess.   With this improvement,
the stem is simply passed in by the side of the
sound.   This is withdrawn, the disk is pushed up
to the projection, which is engaged in the recess,
slightly turned, and the stem is in place with very
little comparative trouble.   This device has saved
weeks of effort.   I regard it as one of my best
inventions.

If desired, the stem of the pessary may be
curved like the uterine sound.   After the stem
has been applied, and sufficient time has elapsed
for the irritation occasioned by the exploration and
application to subside (which generally lasts a few
hours), *if the signs continue, the instrument should
be removed* by the patient, and applied again at a
subsequent time by the physician, until satisfied it
cannot be borne.   The mechanical treatment should
be dropped for a while.   If the suffering should con-

tinue, iodoform or morphia may be applied to the internal surface of the uterus as sedatives. This is best done by Taliaferro cloth-tents soaked in a solution of the morphia, or in a melted mixture of iodoform or cocoa butter. These tents are made out of a cotton bandage, half an inch in width, rolled tightly over a small wire in the shape of an ordinary tent; when coated with cocoa butter, they are easily passed into the uterus by the side of the uterine sound, which is then withdrawn, and the medicated tent withdrawn after a few hours have elapsed. The profession is greatly indebted to Dr. Taliaferro for his simple and effective invention.

Remarks. — Theoretically, this stem pessary is the most perfect of the set. Its axes run in the direction of the long axes of the normal vagina and uterus. It occupies the least space, and allows of almost complete transverse contraction of the vaginal walls, as one cannot imagine any distension being caused by the introduction of a cylinder one-fourth of an inch in diameter. The disk being only as large as the neck of the uterus, it is impossible to conceive of distension from it. It gives efficient support.

Instead of a flat disk, it may be cup-shaped, if desired.

Objection to Stem Pessaries. — The writer is well aware that he is treading upon delicate ground. There is great aversion to the use of stem pessaries on the part of the profession, from the fact that they are liable to cause great trouble by their mechanical irritation. This feeling influenced the writer so much, that for some years he was afraid to employ stems at all. But in 1862, during a visit to Edinburgh, he was made acquainted with the galvanic stem pessary of the late Sir J. Y. Simpson, by his now distinguished nephew and successor. This pessary (a misnomer, by the way, as it has nothing to do with sustaining a dislocated womb), consists of a stem two and a quarter inches in length, and about one-eighth inch and over in diameter. One inch and one-eighth of the point of the stem is made of cylinder of zinc. The remainder is made of copper. The stem is terminated by an oval copper bulb. In the centre of this bulb is a foramen, opposite to and continuous with the inside of the stem, for the purpose of introduction by receiving the point of a uterine sound. It is most successfully employed in cases of amenorrhœa

of an atonic kind. *Its object is to produce irritation, and thus promote the establishment of the menstrual flow.* I was assured of its being worn for
weeks without trouble. On my return to this
country, I tested the use of the instrument thoroughly, and likewise found no trouble. I then
reasoned that, if a compound stem of copper and
zinc (a galvanic battery), constantly irritating by a
galvanic current (when excited by the secretions of
the uterus), could be borne inside the uterus for
weeks, without exciting inflammatory action, why
not an intra-uterine stem of hard rubber,— one of
the most unirritating substances known ? Without
hesitation, the stem was then adapted to the T pessary, and tested with success. I ' have seen my
stem pessary worn for several months — a year
even — at a time continuously, with comfort and
no trouble ; so that they who unqualifiedly con-
demn the use of the stem pessary err, as some
cases *do* tolerate the stem well. On the other
hand, the writer has found cases that did not tolerate it ; and, as it cannot be foretold which will
tolerate, its use must be a matter of experiment, —
legitimate experiment, — as the instrument is so
mechanically correct. An incautious and careless

7

physician, who will not be attentive and painstaking, should never attempt to use a stem or any other pessary. Indeed, he should never practise medicine or surgery, — such a one has no right to interfere with sickness or disease ; but intelligent, prompt, judicious, and observing physicians may, I think, approach the use of the stem pessary without fear of injury. I have had cases where the uterus would expand and swallow up the disk, and I have observed the fundus to bend over the other end of the stem. The difficulty was obviated by a larger disk in the first case, and a longer stem in the second instance, and no mischief followed.

Had I been careless and incautious, there might have been severe trouble : and would I have been justified in condemning the stem pessary instead of myself as the cause? The abuse of a thing is no justification for the prohibition of its use. I am in the habit of saying to my patients that the use of water is necessary to the existence of mankind ; but that, if a person were to be held under water for the space of half an hour, the effects would be any thing but sanitary. So here I would say, Look out for the flags of distress ! IF *there is discomfort, something must be wrong, and the stem should be examined, removed, and adjusted according to the circumstances.*

It is very evident that retroflexion must be dealt with from a mechanical standpoint, — like a contracted joint, or a club-foot, or strabismus. Indeed, it is a question in my mind, whether some cases of flexion may not be caused by a want of muscular tone, or paralysis of the convex uterine wall, thus losing its antagonism to the opposite side, and being pulled out of shape, — as in the contracted joint the flexor muscles overcome the antagonism of the extensors ; or, as in strabismus, a healthy rectus muscle will stretch out its paralyzed antagonized rectus and cause distortion ; or, if the right shroud of a vessel's mast is loosened, and the left shroud tightened correspondingly, the effect will be to set the mast awry to the left.

The uterus is mainly a muscular organ, not unlike the heart, which has powerful symmetrical and rhythmical contractions. We are familiar with similar contractions in the impregnated uterus at the time of labor. But does not the unimpregnated uterus have the power of contraction ? Put a sound into one, and how often we find it expelled into the vagina, unless it is held ! Indeed, I do not recollect of an instance where it has not been expelled, when let go. This fact of expulsion is therefore insisted upon as a conclusive proof that the muscular fibres of an unimpregnated uterus are active and contractile. If this is so, then the morphology is dependent, not upon conformation, as a bone or kidney, — to be changed physically (not physiologically) by forces acting alone from without, — but also by interstitial forces of its own muscular tissues. If the contractility and tonicity of these muscular fibres are normal, and normally antagonized, there results a normally-shaped uterus; but if, for instance, the fibres only of the anterior wall are paralyzed, what can be expected but that the muscular fibres of the posterior wall, — preserving *their* normal contractility as long as they have an antagonism, — will perform their function, contract closely down, and, pulling over the anterior wall, cause the deviation we call retroflexion? If we accept this explanation in talipes and strabismus, why not in uterine flexion?

Fig. 17.

Actuated by these reasons, the writer has suggested the propriety of faradizing or galvanizing the weakened posterior uterine walls, and offers Fig. 17 as an electrode for applying the battery currents to the convex wall in uterine flexions. In this instrument, the posterior half, longitudinally, is protected, while the anterior half is exposed; so that a current passed through will directly traverse and affect the anterior wall, when the electrode is passed into the uterus, and connected with the battery. May not tone and contraction to the convex wall be aided and assisted in connection with mechanical treatment?

In cases of retroflexion combined with vaginismus and irritable uterus, resort may be had to those procedures which have been fully detailed under the head of retroversion.

RETROVERSION AND RETRO-FLEXION COMBINED.

Fig. 18. Here the vagina is so much relaxed that the retroflexed uterus is thrown down as much as the sacrum and rectum will allow. It is common to find this combination. It interferes with defecation more than the disconnected varieties. It is more difficult to treat; but the principles and procedures are the same as already indicated, except, perhaps, in some old cases, where it is found that the uterine sound causes a great deal of subsequent irritation and discomfort, which is not relieved by depletion, or where the indications are strongly against much local depletion (that is, long-standing, profuse menorrhagia, great general debility and depletion, followed by considerable nervous disturbance). In these cases of retroversio-flexion, the patient may be placed on a single inclined plane (a bed, raised

at the foot, or Fig. 12, will answer), and then an
accurately-fitting retroversion loop pessary may be
fitted and applied, with the hope that the flexion

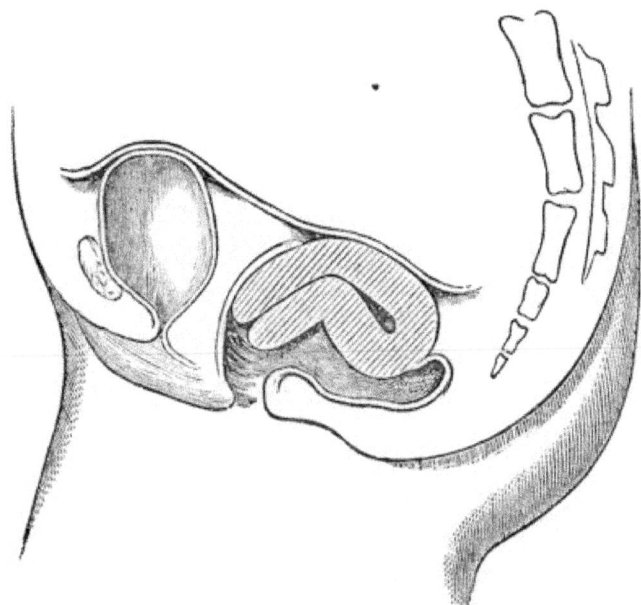

Fig. 18. — Retroversion and Retroflexion.

and version may be gradually removed by the
graduated pressure of the pessary, assisted by the
atmospheric pressure and the drawing downwards
of the abdominal viscera, whenever the shoulders
are placed below the level of the pelvis. In such
cases, the length of time and the frequency of the
application of the pessary may be regulated by the

patient herself. As the case progresses, longer pessaries must be employed ; and their necessity may be indicated by the hook pressing against the perineum, and by the vaginal sound. The relief afforded by a simple support of the retroverto-flexed uterus in its abnormal bed, or a little above it, in these old cases may be all that we should endeavor to obtain. The changes are so permanent and fixed as to be incapable of complete restoration ; and it is better to be content with a partial restoration, rather than by a complete one to make the patient worse. The abnormal conditions having become a second nature, and the system having accommodated itself to them, it is idle to expect relief from a complete restoration. The human system is subject to the laws of habit. Neuralgic pains and nervous disturbances become habitual like other things, after their causes have operated for years. Remove those causes, and not always do the effects also disappear. It is so with physics : the boy who starts a snowball down a mountain-side, producing an avalanche, ceases to act the instant his hand gave the push ; but the results of that push are just as effective as if it had been possible for the initial force to have been

· maintained throughout the whole period of descent. Why, then, it may be asked, if you cannot *cure*, interfere? Because, as the venerable Dr. James Jackson once wrote, "curing" means "caring" (*cura* = a care). If we relieve our patient of half of her distress, we are "curing" her, and exercising legitimately the functions of our art in a beneficent direction. And if she, by wearing a properly-fitted retroversion pessary, can support her womb so that locomotion can be painlessly performed, this, also, is an effort which should be assisted and encouraged by the humanity of the profession.

ANTEVERSION.

Definition. — Anteversion (Fig. 19) is an abnormal condition of the uterus in which it is displaced forwards, without bending its long axis. It is the reverse of retroversion. The fundus is turned forward, and rests on the bladder, compressing it antero-posteriorly, and causing in general symptoms of vesical and urethral irritation. The os points backward, into the concavity of the sacrum.

Diagnosis. — The forefinger, passed into the vagina, comes into contact with a flat, hardish body, forming the ceiling of the vaginal cavity. This is the anterior surface of the uterus. There is no anterior-utero vaginal *cul-de-sac*, nor posterior, — both being obliterated by the malposition. The vaginal diameters are reversed, as in retroversion. The uterine sound can be introduced only by bending its point, turning the concavity downwards; placing the handle forward, pass the point over

the finger to the os ; engage it, and, drawing the
handle of the sound backwards to beyond the
thighs behind, gently slide the instrument into

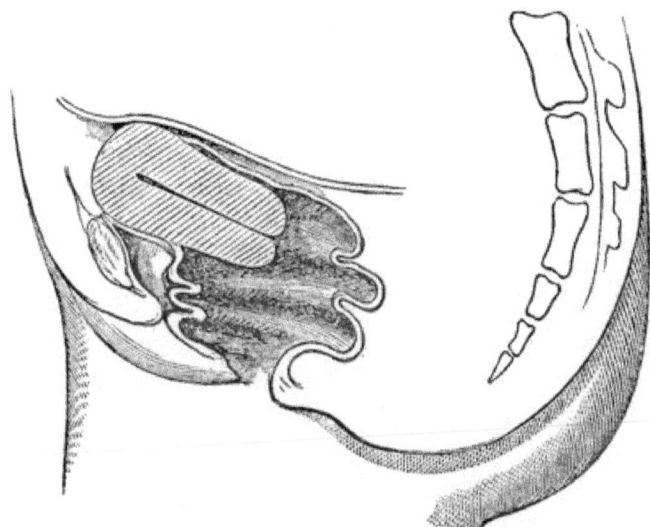

Fig. 19. — Anteversion.

the cavity.   The degree of obliquity varies.   As
the uterus is naturally anteverted (*vide* Fig. 1),
the deviation of an abnormally anteverted uterus
is less than that in retroversion.   The bladder and
the pubis prevent it, also, from being so complete
as its opposite version.   Care should be taken not
to confound any other disease with this displacement, or not to overlook any complication of metritic, intermetritic, or perimetritic disease.   The

uterine sound is the best diagnosticator. Sometimes a uterus may not appear, on digital examination, to be anteverted, until after the uterine sound
has been introduced; and then the physical changes
brought about will clearly show what the difference is.

Taking an old case, with a flabby, inert uterus,
the explorator will occasionally find much difficulty
in passing the sound. The uterus forming the
dome of the spherical or globular vaginal cavity,
— lying partially transversely across, — the moment the sound is entered into the os, the womb
will rise up from the vulva; and, if the air gets in
from the outside, the vagina will be blown up into
a rigid, inelastic, hollow ball, the uterus forming
the upper segment of that ball, so that it is impossible to enter the sound at all. It cannot be bent
to the curve of the uterine axis and yet be capable
of introduction and insertion. If you keep trying
for a long time to reduce the elevated organ, you
may induce inflammation. Desist from your efforts
and let out the air; the os will descend within
reach, and it seems now as if it might be easily introduced. It is tried, but the uterus rises with the
sound (the point probably pocketed), the atmos-

phere rushes in, and up it ascends into its former elevation and rigidity. The better and only way is, if your patient will allow it, and is not wholly discouraged, to wait a day or two, till the organ may have recovered tone enough to resist the sound, and not allow it to pocket itself.

I have lately seen the inflation of the vagina by the atmosphere recommended as a means of putting the uterus into its proper place ! The patient is instructed to lie down with her shoulders lower than her pelvis, then to introduce a small tube into the vagina, and allow the air readily to enter, which it certainly will do. But how this procedure is to cure the anteversion, for instance, no one who has tried to replace the uterus under these circumstances will understand. And then, besides, the transverse vaginal sphincter muscle is distended with a pressure nearly equal to fifteen pounds to the square inch! But we have shown that vaginal distention is the essence of the disease. Now, can the suggester propose to cure the disease, not by a like one, but by making it worse? It appears to the writer a very unreasonable proposal, and liable to do harm by admitting air on to a mucus membrane, and by stretching the vagina by such a pow-

erful, though evenly-applied, force of over one
hundred pounds. I sincerely hope it will have but
few adopters.

Subjective Symptoms of Anteversion. — Generally, the flags of distress of the nervous system are
out all over the body, or localized about some single spot ; as in the case of the backward versions
and flexions, with the additional one of irritable
bladder. Micturition is often (not always) frequent and urgent. This is readily explained by
the uterine fundus pressing or crowding upon the
urinary bladder, so that its capacity is diminished ;
and hence the calls to empty it are more frequent.
Besides, the bladder is more irritable from this
pressure, and from the determination of blood to the
pelvis and its contents. Often, too, this is combined with ovarian soreness and irritability ; the
bowels are sore sometimes : it hurts to walk about,
cough, lift, run, or strain. Leucorrhœa is generally
present.

Frequency. — I am inclined to think it is the most
frequent kind of displacement. It sometimes happens after a retroversion, or rather the retroversion
is sometimes changed into anteversion. Owing to
this frequency, it is of more importance to have the

subject fully studied here; the more so, as Dr. Thomas, of New York City, — the most eminent living authority on these subjects, — declared, in June, 1870, that, before the introduction of the pessaries described here, there was no cure for anteversion, but that now he treated it with the same facility and success as retroversion. I may, perhaps, be pardoned in mentioning this, as any inventor feels well satisfied for his expenditure of pains and trouble (if he ignores pecuniary considerations), when a competent person publicly announces that his devices fill a void not filled before, relieve suffering not relieved before, and cure the hitherto incurable. The same gentleman wrote: "The public, the profession, and suffering humanity are under the greatest obligation to you for the invention of your pessaries." As I have freely given up my right as inventor to the profession, this response is very gratifying, — the more, because I have derived the most insignificant pecuniary reward for my invention. The writer again asks pardon for introducing these words here; but it does seem that the attitude of the medical profession as to rewarding the efforts of its inventors is not calculated to encourage them. Invention needs patronage. Financial and inventive genius are rarely found allied. Original observation needs patronage. The exhaustion of the dynamic vital forces in perfecting new mechanism or conducting new observation is great. Were large sums of money devoted to the reward and support of óriginal investigation and invention in the departments of medicine and body mechanics, there is reason to believe that efforts in this direction would be stimulated as in other departments of human effort ; like that, for instance, of the weather reports. Aged physicians who have acquired a competency owe it to themselves and the public to give up their practice to the many that need it, and devote their remaining days either in pursuing original investigations themselves, or setting others to work at it ; taking those points which their long and laborious lives have shown them to need it the most. If they do not have the means themselves, they might collect them and see to their disbursement. The wealthy public would be glad to assist enterprises inaugurated under such auspices ; for the age, character, and standing of the supervising physicians would be a sufficient guaranty to any one. If such physicians would do this, they would erect to themselves a nobler monument than art could make of brass or marble.* For the idea of the adaptation of the retroversion pessary, I am indebted to Dr. Thomas, of New York, though he magnanimously disclaims it.

Indications of Treatment: —

(*a*) Replacement by uterine sound.

(*b*) Holding the womb *in situ* with the sound.

(*c*) Then, by vaginometry, select some physical means to keep the vagina extended in its long, normal diameter.

(*d*) Support the uterus in the anterior *cul-de-sac.*

(*e*) Allow contraction of the transverse vaginal fibres.

(*f*) Cause the apparatus to be manageable by the patient.

It must be remembered that the uterus is naturally inclined forward ; hence, the supporting agent must rest in the anterior *cul-de-sac* without pushing the womb up so far as to push it over and make it retrovert, or even raise it beyond the line of normal inclination forward.

* See letter to the Middlesex East (Mass.) District Medical Society, — " Boston Medical and Surgical Journal," December, 1874.

In carrying out the indications for anteversion,
I have relied upon the retroversion loop, or T,
pessaries, modified so as to have the loop bent
forward, or the bar of the T reversed, so that its
concavity looks backward.

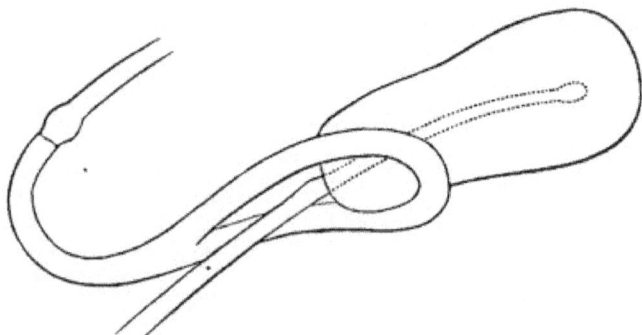

Fig 20. — Anteversion Loop Pessary applied to Uterus.

The process of application is the same as for that
of retroversion, only substituting front for back.
However, as the subject is of so much importance,
it will be described at the risk of repetition.

(*a*) Having properly diagnosticated the case and
prepared it (as in retroversion), the uterus is replaced by the uterine sound.

(*b*) It is then held *in situ naturali* by the
sound ; and then

(*c*) The vaginal sound is passed into the ante-

rior *cul-de-sac* over the left forefinger, previously
introduced; or the forefinger may be passed in
subsequently, and the *cul-de-sac* examined to see
if the bar of the vaginal sound is properly placed.
This being the case, the point where the perineum
touches the measuring sound is noted, and the
sound withdrawn.

(*d*) An anteversion loop pessary is selected one-
half inch longer than the ascertained measurement.

(*e*) The handle of the uterine sound is then held
by an assistant; or the patient raises herself on her
left arm, reaches with the right arm over the right
thigh, and holds the sound herself. I am obliged
to ·caution my patients against pushing in the
sound too nard, but instruct them to hold it lightly
and firmly.

(*f*) The pessary is then passed over the uterine
sound to the patient's hand. The physician then
seizes the handle. The patient lets go her hold.
The pessary is passed up against the vulva. The
patient next takes the handle of the sound again.
The physician lets go his hold. Then turning the
convexity of the loop forward, assisted by both his
hands, the physician introduces the pessary into
the vagina, and it generally slides into place in the

anterior *cul-de-sac.* (*Vide* Fig. 20.) A digital examination should then be instituted, to see if the
loop is actually in its proper place. If so, the
sound is withdrawn, while a hold is maintained
upon the suspensory cord. The physician should
then examine the distance between the pessary and
the perineum, and if it is one-half inch, it is right;
if not, it should be readjusted, and another pessary
applied, if necessary. Retaining the hold upon
the cord, he gives his right hand to that of the
patient, and then gently raises her off her support,
up on to her feet. The belt is passed around the
waist, and fastened so as to give the right tension
to the cord, which should be neither too loose nor
too tight, — a medium tension. If the cord is too
long, shorten it. The patient should be requested
to walk across the apartment a few times, and sit
down on the bed first; then upon a harder support, as a chair. If the pessary feels comfortable
and does not hurt, she may be allowed to wear it
until it gives her trouble; *then remove it.* The
physician, at his next visit, may go over the ground
once more, and correct any errors. He will probably find the uterus anteverted again as soon as
the pessary is removed. This cannot be helped;

it is due to the long-standing of the disease. A uterus of a few days' or weeks' displacement will stay up where it belongs, after replacement with the sound, possibly with no pessary. When the vaginal sphincters and the ligaments have been relaxed, can their normal tone be renewed by a few hours of relief? Moreover, in anteversion, the round ligaments are probably contracted by the long relaxation of their normal tension ; if so, they will tend to antevert the uterus at every opportunity. · For these and other reasons, the patient should be instructed how to introduce the anteversion pessary ; fortunately, it is a very easy matter.

Directions for replacing an Anteversion Pessary by the patient herself. — Lie upon the back, shoulders raised. Take the pessary, introduce the loop within the vagina, and gently push into place. The convex edge of the loop follows the short anterior vaginal wall so easily, that it is almost impossible to get the instrument in wrong. A hold should be had upon the cord, and the belt then adjusted around the waist.

Thomas ("Diseases of Women," p. 370) says: "A very simple one of the former kind is a modification of Cutter's retroversion pessary. The

upper extremity of this form of Cutter's pessary
has a bulb attached to it, and is so bent forward
as to strike the base of the bladder, anterior to
the cervix. This is introduced by the practitioner,
and its method of introduction and removal fully
explained to the patient. She is instructed to
remove it every night upon retiring, and replace
it before rising in the morning. By it the cervix is pulled forward, the utero-sacral ligaments
stretched, a tolerance of the foreign body estab-
lished, and a pouch, or pocket, created anterior to
the cervix, which will accommodate in time the
anterior bow of the pessary (Fig. 21). *The bulb
pessary with 'external attach-
ment may in any case be used
as preparatory to an internal
instrument.* After the former
has been used for a month or
so, the latter will be generally
applicable. One having experience with these two instruments can almost always
tell, without experimentation,
which will be appropriate. If
there be a pouch anterior to

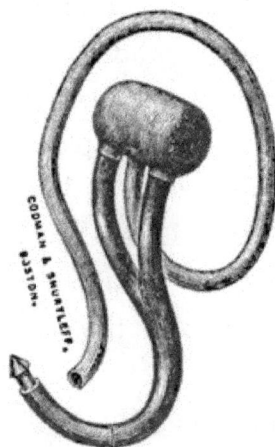

Fig. 21. — Thomas's modified
Anteversion Pessary. Bar inflatable.

the cervix when the base of the bladder is pressed up by the finger, the internal pessary will be tolerated. If there be none, and the tissues resist pressure by the finger, it cannot be employed until space has been created by the other instrument."

Remarks. — In these cases, I would employ the stem pessary just described.

Fig. 21 is a device rendering the bar soft and inflatable by air or water, to give a yielding and elastic cushion. The cut is an exact picture of Thomas's anteversion pessary, except the rubber tubing.

Fig. 22 is a filled anteversion loop pessary. It may be made of soft rubber ; turned back, it would answer for retroversion. This worked very well in the case for which it was designed. It is not so accurate as the filled T pessaries.

Anteversion T Pessary (Fig. 23). The same general instructions may be given for the selection and ap-

Fig. 22. — Filled Loop Pessary.

plication of the anteversion T as for the anteversion loop pessary.

Sometimes there is no anterior or posterior utero-

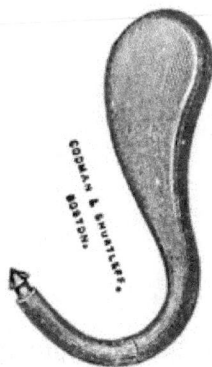

vaginal *cul-de-sac*, — the vagina being united to the uterine neck just at the os. This may have been the normal condition, — my limited experience does not substantiate this position. In such cases, the

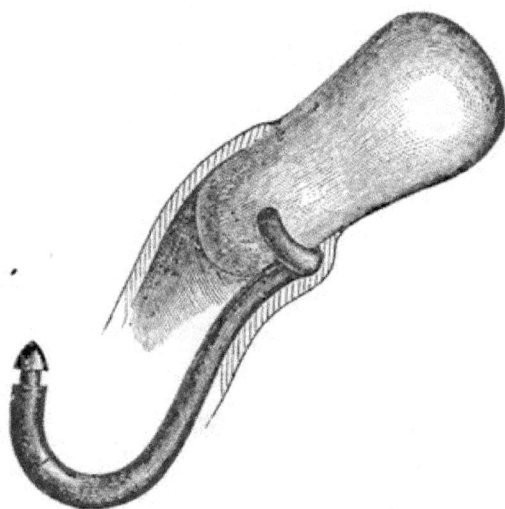

Fig. 23. — Anteversion T applied to Uterus.

maintenance is effected by means of the stem pessary. The method of application is similar to that already laid down under the heading of retro-version. The same care in its employment is necessary.

Remarks. — The great objection to the use of this form of pessary lies in the fact, that the

anterior vaginal wall and neighborhood are more
sensitive than the posterior wall ; hence the greater
liability to irritation from the contact of a foreign
body. And there are cases which cannot tolerate
any thing. These should be treated very cautiously,
or let alone. In all cases the tolerance should be
tested by trial, and the instrument withdrawn if
there is trouble. Time should be allowed for the
irritation to subside ; and then the physician, or
patient, may introduce it again. In this manner,
the parts may be gradually accustomed to the instrument, as the mouth becomes accustomed to a
new set of artificial teeth. The procedure during
defecation is like that described for retroversion ;
or the instrument may be removed before, and replaced after, the act.

Fig. 24. In this displacement, the uterus is bent
forward without removing the lower portion out
of the line of the vaginal axis. The touch is the
best means of diagnosis. The os is found sunken,
but in the middle of the vagina. The fundus is
felt behind the pubis, making with the remaining
part of the uterus an angle, the *point* of which
varies from a position close to the os, to one which
is within three quarters of an inch of the fundus.
The size of the angle varies, also. The next step
in the diagnosis consists in introducing the uterine
sound, generally a difficult and tedious operation.
The sound should be bent to correspond with the
angle of flexion, and in its passage should be
assisted by elevating the fundus with the fingers,
and sometimes the vaginal sound, inside the vagina ; and also by pressure over the pubis outside.
The degree of success with the sound is an important feature in the diagnosis. If the uterus is
completely penetrated and found to be mobile, we
infer that the flexion is the main element of the dis-

ease. If not, we should explore carefully for some other accompanying pathological condition which may be the main disease, and take the case out of

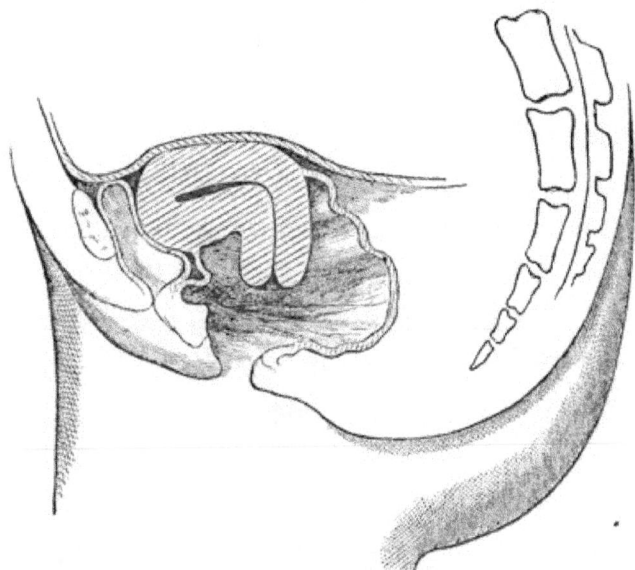

Fig. 24. — Anteflexion.

the class treatable for displacement. If the flexion is uncomplicated, and is readily retained in place by the sound, then the vaginometry should be employed, and a loop or T pessary fitted as in anteversion. But if there be a disposition for the reduced flexion to return, then I know of no mechanical treatment but that of the stem. Its introduction and management have been described under retroflexion.

ANTEVERSION AND FLEXION
COMBINED.

These (Fig. 25) constitute a formidable combination against successful treatment. The uterus forms a portion of the vaginal dome, reaching from the pubis in front, upwards, to beyond the middle of the highest point of the vagina. The operator will at times find it very difficult to bring the organ *in situ naturali*, as every thing is so much out of line, especially when the uterus is flabby and yielding. The sound, bend it as much as you may, will "pocket" itself in the upper wall, and will not advance further, but carry the womb upwards, deceiving the careless. How do we know when the sound has properly entered? By the position of the uterus being in the right direction. With a well-defined *cul-de-sac* in front and behind, the sound is more loosely engaged in the uterus than when packed into a pocket or fold. A feeling, also, which it is difficult to describe, is produced, and which expe-

rience will teach. The proper course lies in waiting for the uterus to become more tonic by general
treatment; or to employ a flexible sound, and then

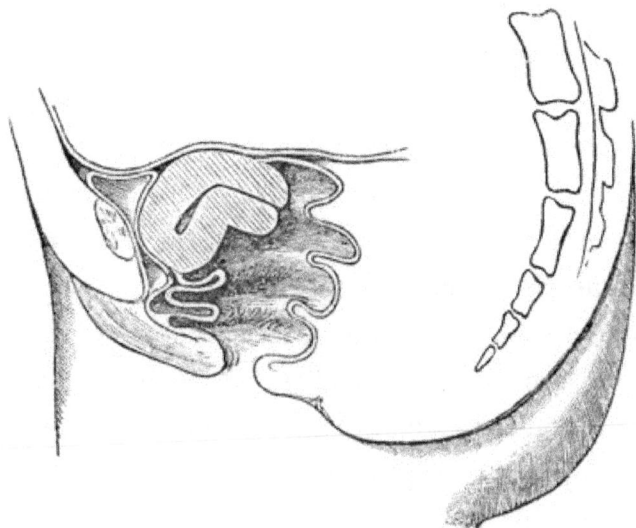

Fig. 25.— Anteversion and Flexion.

follow with a stiff sound. The other procedures
of the management are the same as laid down
under anteflexion. Care should be taken to ascertain the complications. If the uterus is hypertrophied (and if we repeat ourselves here, it is
hoped that it will be overlooked, as the subject is
so important), remedies should first be applied to
reduce the size of the organ. Depletion is one of
the promptest and most successful agents to bring

the organ down to normal size. Two or three leeches I have seen to work wonders in reduction. Where blood-vessels have been distended by chronic congestion for months and years, it is fair to infer that the vascular walls have become weakened by the over and long-continued distension, so that the stimulus of chemical and galvanic agents is *nil* (as experience shows) to cause a normal contraction. Indeed, it seems as if the contractility were nearly gone. So that what more sensible or common-sense procedure can be adopted, than to let out the blood by mechanical means, and allow the distended vascular walls to collapse, if not contract, by sheer want of contents? These procedures have proved to be the most efficient curative agents in old cases of throat-disease. We are aware that this doctrine is contrary to that of the present time; but is it wise blindly to follow fashions in medicine and surgery as we do in dress? Intemperance in depletion is no argument against its *temperate use.* To decry and wholly refuse to employ a therapeutical agent or process, because it has been abused, is not a temperate or sensible position for a wise physician to take. In this respect, the medical profession yields too much to popular sentiment, and there is reason to believe that loss of life ensues simply from the want of a judicious letting of blood. If there is chronic heat, tenderness, throbbing, with enlargement of the uterus or vagina, or both, in almost any case of uterine flexion or version, one will rarely err in applying scarification or leeching to the parts affected. It prepares the way for topical and mechanical treatment, by the simple unloading of the tissues of the superfluous, distending blood.*

The account of this part of the subject would be incomplete, if all mention were to be omitted of some peculiar conditions that give trouble in the treatment of anteversions and flexions; *viz.*, the irritable intra-uterine ulcer, and the doubling of the uterus over the pessary. The first case is diagnosticated by the uterine sound : when it passes in, never so gently, it produces the most exquisite anguish, which ceases as it passes by, even up to the fundus, and is renewed as the point of the sound passes the irritable ulcer in going out. Leeches have been used in the uterus and over the abdomen just

* As Prof. Hodge pointed out, an enlarged and tender uterus is not always *inflamed*, but *irritable* from displacement, and is relieved by reposition. It is well to remember this, and give the case a trial without depletion.

inside the anterior-superior spinous processes of the ischia, — as it is a remarkable fact that the ovaries, one or both, are sensitive and sore in these cases of irritable intra-uterine ulcer. The ulcer causes the ovarian pain by sympathy, probably. Next, morphia and iodoform may be applied by Talia-ferro's tents. A stem pessary may then be used on the principle of dilatation. If the Cutter stem is difficult to bear, I have found it advantageous to employ an *ordinary* stem ; then, introducing a

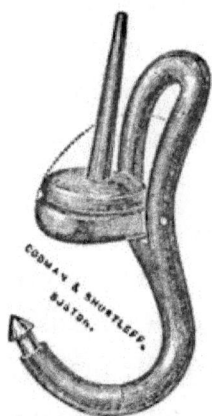

Fig. 26. — Bracket Pessary.

bracket pessary (Fig. 26), hold the stem in place without the rigid hold of the first-named stem. Something, too, is gained by the support in front of the bracket pessary. This bracket pessary is like the ordinary anteversion loop, except that a rudely triangular piece of hard rubber is articulated on ,to the narrow end of the fenestra in such a manner as that, when placed at a right angle to the pessary, it strikes a projection and remains fixed. It then supports the disk of the stem. In the cases for which it was invented, the ordinary stem was not borne, but the combination of the bracket and stem pessaries was comfortable. The ordinary anteversion pessary will sometimes support a common stem very well in tolerant cases, but not in the sensitive ones that happen with intra-uterine complication.

The other complication of bending of the womb over the end of the loop is a homely way of expressing a trouble which the writer has met with. The pessary will be fitted accurately before the uterine sound is removed, so that there is no doubt that the placement is correct; but on a subsequent visit, although great relief has followed the use of the instrument, still the digit finds that the uterus is bent over partly double, and no other physical reason can be offered but that the loop has slipped backwards out of the *cul-de-sac*, and become saddled by the uterus straddling it in front and behind. This is probably caused by the distension of the vagina being so great that the loop has not a point of resistance strong enough to hold it. This may be met by putting in a Cutter stem pessary, or, as just mentioned, the common stem and the bracket anteversion loop pessary, until the vaginal sphincters may have regained their tone. Should this not succeed, I should give up the case.

Mobile Uterus. — These are cases in which the organ moves very readily, either backwards or forwards. It is found in one malposition at one visit, and at a subsequent visit in yet another. If replaced from retroversion, it will go over to anteversion ; and *vice versa.* The ligaments appear to be completely relaxed, and allow this wabbling about. In such cases, I have preferred to convert the ante into a retro-version, and treat them as such ; allowing the organ to be slightly turned back, so as to be less liable to be thrown forwards during the body movements.

LATERO–VERSION.

Two varieties, right and left. This is found associated with prolapsus, and sometimes with retroflexion. It is not common. In the left variety, the os is found turned toward the right side of the lower portion of the vagina; sometimes it presses upon the floor of the pelvis, and the fundus is turned towards the right. It is found necessary, in entering the uterine sound, to bring the handle over the left thigh (the patient's position being on the left side, thighs flexed, legs semi-flexed), and to assist by the left forefinger. When the sound is entered, the direction of the displacement is shown by the course of the sound. The organ should then be replaced, and the vagina measured by the sound. A stem pessary may be selected and applied. If this is found not suitable, either the loop or retroversion T may be resorted to. The extension of the vagina to its normal length and axes produces a tension of its insertion into the uterus, and the counter tension causes the erection of the uterus into a right line.

The right variety of latero-version is similarly detected as the left, reversing the description.

Latero-flexion. — In this deviation, the uterus is bent sidewise. It is rare and difficult of detection. The uterine continuity must be traced from the os upwards by a flexible sound. The treatment advised is the replacement and adaptation of a Cutter stem pessary.

PROLAPSUS.

In this displacement, the uterus descends downward in the direction of the long axis of the vagina. The amount of descent varies from the lowering of an inch to a complete exit outside the vulva. The vagina is reflected in the same direction; it is turned inside out, — intussuscepted. The upper part becomes the lower, and the fundus of the womb is covered and concealed within the vagina. The orient point of diagnosis is the os uteri. It is doubtful whether this description of uncomplicated prolapsus is common. On the other hand, prolapsus combined with version and flexions is very common. Judging from medical literature, this displacement is the most frequently met with.

Treatment. — When the uterus can be replaced within the pelvis near its normal site, by pushing in, the indication is to hold it there. If the uterus

is too much hypertrophied to be returned, measures
for reducing the size must be adopted. Depletion
by leeches may be resorted to, — the patient being
kept in bed with the head lowered on a single in-

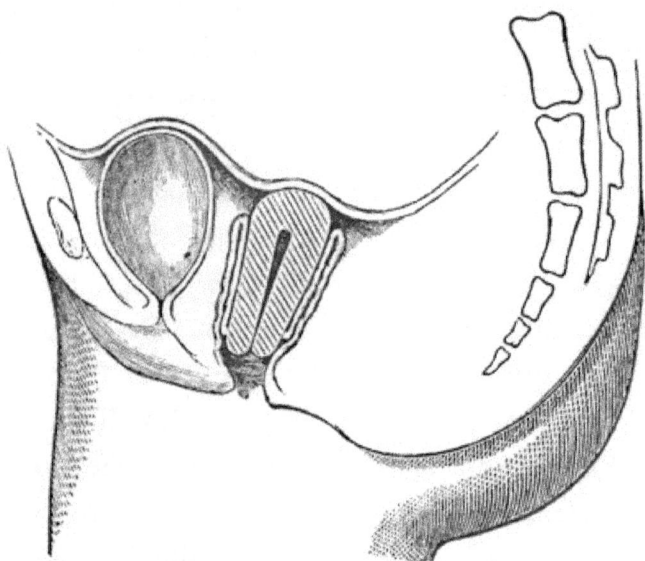

Fig. 27. — Prolapsus.

clined plane. My resting chair will answer for this
purpose very well. The vaginometer may be used
to get the diameter of the uterine neck before replacement. After replacement, the vaginal sound
will give the distance of the os from the perineum.
A pessary one-half inch longer than the measurement, and provided with a ring (Fig. 28) which has

exactly the diameter of the uterine
neck, is selected and applied to the
organ, and left there. The ring pessary is a modification of the T
pessary before described, only that
the bar is continued into a complete
circle. If the womb is rather sensitive and heavy, it may happen that
the bearing on the ring is too small ;
if desired, it may be substituted by

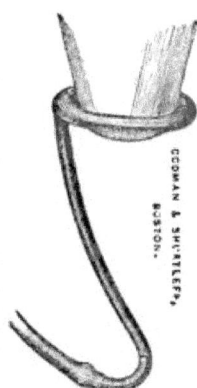

Fig. 28. — Ring Pes-
sary.

the cup (Cutter's) pessary (Fig. 29). This cup is
merely an expansion downwards and inwards of the
under surface of the ring pessary. It is perforated
for the transmission of the menstrual
and leucorrhœal discharges. The
angle of the ring and cup with the
cylinder of the pessary should be
less than a right angle, as the
natural inclination of the uterus is
forwards. In this manner the
uterus is supported, and the dilated and loosened vagina is allowed
to contract, and thus rendered

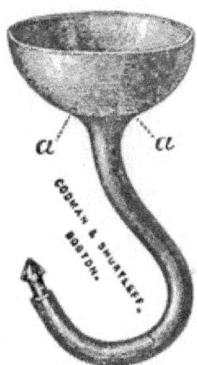

Fig. 29. — Cup Pes-
sary. *a a* — Voram-
ina.

more liable to return to its normal condition of
columnar support. The vaginal contraction may

be further promoted by a sensible suggestion of a
physician whose name has escaped me ; namely, of
packing the vagina with dry tannin, after the pessary is replaced, thus pricking up the mucus membrane.   The tannin, in some cases, will irritate the
mucus membrane, and make it very sore, and
cannot be employed further.

The patient with prolapsus may assist the introduction of the pessary by inflating the vagina by
atmospheric pressure, as already alluded to.   It
will greatly assist in the elevation of the uterus.
Indeed, some uteri cannot be reduced without lowering the patient's head to the floor.*

* Dr. Thomas — "Diseases of Women," fourth edition,
Philadelphia, 1874, p. 348 — writes : "No pessary with which I
am acquainted so universally answers the indications of supple-
menting the actions of the utero-sacral ligaments and sustaining
the prolapsed vagina, rectum, and bladder, as Cutter's admirable
pessary."

PREVALENCE OF VERSIONS AND FLEXIONS OF THE UNIMPREGNATED UTERUS.

An old physician, of very extensive practice in the country, once remarked that, supposing women everywhere should be subjected to a physical exploration of the vagina, in his opinion, *nearly all* would be found with some abnormal uterine deviation. There is no doubt this assertion is true ; and we must admit that it is possible for considerable — nay, enormous — deviations to occur without exciting subjective symptoms of sufficient importance to attract attention or warrant interference. The late venerable Dr. Hodge, of Philadelphia, used to relate, in his lectures, a case of a market-woman who carried on her business and slept in her wagon the whole year round, yet bore her enlarged and prolapsed womb hanging down to her knees between her thighs ! This extreme example shows what great changes in physical circumstances the system will endure, and gives the physician a sufficient warranty for saying that uterine deviation, in some cases, may be overlooked, ignored, or disregarded. But it furnishes no ground for making a sweeping general conclusion that *all* such cases are like that pitiable, though not disturbed, market-woman. It is equally true that there *are* women in whom versions, flexions, or displacements, singly or combined, — sometimes hardly perceptible, — *do* constitute disease ; and these cases do ask for treatment, and, when falling into the hands of physicians, ought not to be cast away as unfit for treatment.

The subjective symptoms are chiefly neurotic. They are often mimotic. They are not present in the same shape or form in each and every case. One set of nervous disturbances will be present in one case, and another set in another. A case has sometimes its own peculiar phases, differing from any other. No subjective symptom is pathognomonic without being confirmed by objective symptoms disclosed by physical exploration.

Any woman, old or young, before or after the menopause, may be suspected of suffering from uterine deviation, who complains of, —

(*a*) Neuralgic pain or bad feeling in the head, neck (cervical spinous processes), epigastrium, præcordium, hypogastrium, lumbar region, along the course of the sciatic nerve, on the pectoral muscles (sometimes localized in circular spots one inch in diameter), or over the ovaries.

(*b*) Of a feeling as if a string were tied around the waist, or a handkerchief about the neck (stricture).

(*c*) Of inability to carry weight, or promenade, or go up or downstairs from weakness in back.

(*d*) Of difficult defecation.

(*e*) Of difficult and frequent micturition.

(*f*) Of prickling and numbness in the limbs (a very common symptom).

(*g*) Of anomalies of vision without physical cause.

(*h*) Of great nervous irritability.

(*i*) Of syncope simulating epilepsy.

(*j*) Of hysterical convulsions.

(*k*) Of impending insanity.

(*l*) Of palpitation and pain in the heart, without the physical signs of organic lesion.

(*m*) Of leucorrhœal discharges (though this is by no means a constant or reliable symptom), etc.

If the physician, after a careful and thorough examination, is unable to find a satisfactory cause of any or all these symptoms, suspicion should fall upon the uterus. It is a *duty*, then, to explore the vagina ; duty for the physician for accurate diagnosis ; duty for the patient in order to be relieved. For treating any of the symptoms above detailed, without touching their primal cause, is like the conduct of the Dutch Admiral, who, desiring the wind to be always in the one direction that he liked, had his man every morning set his vane, or weather-cock, in the given points of the compass, although the aerial currents might be blowing exactly opposite ! However apocryphal this story may be, it illustrates the folly of treating a case by the symptoms, when it is possible to ascertain the cause of those symptoms. Disease, we well know, is made up of facts we call symptoms ; but the symptoms are only manifestations, not the abnormal condition that has caused the symptoms. The language employed in the New Testament is, we think, clear on this point, where, in speaking of our Saviour, — " He healed all manner of sickness and disease," — " sickness " may be understood to mean functional derangement, and

"disease" to mean organic, pathological changes
of a morphological or interstitial character, either
of the whole body or parts of it. Nothing is
easier than to "*doctor*" symptoms. It is the
basis of true *empiricism.* — To give another illustration. Fatty degeneration is a condition, a
change of pathological character, which consists
in the tissue changing into fat. Now, the symptoms of fatty degeneration vary exceedingly, according to the organ invaded. It forms the essence,
if we are correctly informed, of cataract in the
eye; the basis of Bright's disease of the kidneys;
the cause of rupture of the heart; the cause of
atheroma, and hence softening of the brain; of
cerebral hemorrhage, which produces hemiplegia;
of angina pectoris; and the subinvolution of the
uterus after labor. Here certainly there are, at
least, five different diseases with a physiological
process all marked by the most diverse symptoms,
and *yet characterized and caused by the same lesion
of histological elements in all.* To heal these complaints, one must change the diseased structures
involved.

The vaginal examination throwing light upon
the true condition, the diagnosis by exclusion is

in most cases complete. It is, then, idle to treat
the neurotic disease as the chief trouble, when, to
repeat a simile, it is nothing but a flag of distress that Nature hangs out to show that there is
serious trouble somewhere crying for relief. There
are counterfeits in disease as well as in money
and goods. Professor Hodge, before referred to,
used to say that there was hardly any neurotic
disease which was not imitated by the symptoms
which resulted from a flexed or verted uterus.
The profession is greatly indebted to him for calling attention, more than twenty-five years ago,
to the doctrine of reflex-irritation in uterine
disease. His valuable work, entitled "Diseases
Peculiar to Women," published by H. C. Lea, in
1860, is very full on this subject, — perhaps too
full; yet none too full, when the quiet apathy of
the profession and the severe sufferings of women
are so prevalent in civilized lands. There is need
of line upon line and precept upon precept in this
matter of reflex-irritation, as it goes farther than
the list of subjective symptoms the writer has detailed, — which are only the most prominent and
common. Among the instances mentioned by my
preceptor, Dr. Hodge, a few are quoted to show

that my own statements are moderate when compared with those of this great master and teacher.

"*Languor.* — Patients who are apparently well, hence objects of criticism, under the idea that the will alone is wanting to dissipate all their ailments. Irritations along particular nerves, hyperæsthesia of the skin, spinal irritation, cerebral disturbances, convulsions, catalepsy, delirium, sudden loss of consciousness and of muscular power, occasionally simulative apoplexy and paralysis, intellectual and moral disturbances, depression of the mind, hallucinations, spasms (which patients call internal spasms) involving the pharynx and larynx, asthma, tired feeling about the chest, cough, aphonia, cardiac disturbances, tumors of the mamma," and so on, — we fear the list is already too startling, and may not be believed. The writer, perhaps, may be allowed to testify that he has seen in his own comparatively limited practice, of about twenty years duration, enough to satisfy him that these statements are correct in the main. They deserve careful study by all interested.

Attitude of the Profession towards Uterine Diseases. — "After all the attention and ingenuity, and even science, which have been directed to this

point in times past and present, no suggestion has
received the general sanction of the profession.
Innumerable as have been these suggestions, each
has but a limited number of supporters. Many physicians have avoided such cases entirely, surrender-
ing them to every variety of empirical experiments ;
so that women remain too often wretched sufferers,
spending days, months, and even years a prey to
disorders which disturb every corporeal function,
and which often pervert the whole intellectual and
spiritual being. There has been no lack of ingenuity, and no want of experiments with more or
less success ; but these efforts have not been generally well directed : the proper position of the
organ in health, its means of support, and the
suitable scientific indications to be kept in view
for the relief of displacements, have not been sufficiently developed. There are, however, intrinsic
difficulties in all mechanical arrangements operating
on vital tissues endowed with sensibility, and in
these cases often terribly exalted. In addition to
this, there is always present the opposition from the
weight of the superincumbent viscera, and the great
pressure from the abdominal walls, under the ever-
varying positions and motions to which the body is

constantly subjected." The profession thus unite in the desirability of maintaining the normal position of the uterus, and are waiting expectantly for some one to introduce an efficient means of accomplishing this result. It is not, then, a question of *desirability*, but of the *how* to do it. It is just here that the writer's humble contribution comes in, as an attempt to solve this problem. The prevalence of uterine displacements demands that physicians should recognize and do what they can to relieve them. There is no doubt but that these difficulties are continually ignored ; and not only this, but those of the profession who advocate the claims of such affections have been treated with derision and contempt. When it is remembered that until very lately the science of gynæcology has not even been pretended to be taught in the curriculum of most of our Medical Schools, is it to be wondered at that the profession have so generally ignored these diseases ? If this method of winking out of sight sufficed for the healing, this course would be right. But is it manly, fair, or consistent for educated medical men to act thus, when in every community women are suffering, with no sympathy, and often ridicule added to their distresses ? No ! one would be false to himself, his profession, and his constituency, if he did not exert all the powers of mind and body to discover and heal, if it may be, complaints which do sap and destroy the maternal head of many families; thus impoverishing and weakening the sources of true civil prosperity. And this is the reason why this *brochure* is offered, in the hope that it may furnish a clew to the right treatment of these disorders. It would be well for those who are not fully acquainted with these subjects, before they decry those who are well informed, to investigate carefully before they go further. Ignorant detraction is very disgraceful; sometimes it is more than this, — even wicked. No one pretends, as some have alleged of gynæcologists, that every woman who is sick has a diseased uterus. The question is, when a more satisfactory reason cannot be found than that certain females are "nervous," "fidgety," and "nothing ails them," and no vaginal exploration has been made by a careful and intelligent observer, whether the case does not demand a thorough physical exploration by her attendant? It is a false delicacy which shrinks from such procedures after an unsatisfactory examination of other regions of the body. Mistakes are made by physicians, said the late Dr. John Ware, more in not being thorough in their examinations than in any thing else.

Case to Show how Retroversion was Ignored. — A virgin, eighteen years of age, fell insensible while walking in the street. The attack simulated hysterical epilepsy. She had had similar attacks before. A physician was called. He appeared to regard the case as one of slight moment, — " as nervous." He administered remedies by the mouth, with the inspiring statement that they would cure her. The anxious mother asked, " Can she walk out any more ? " — " Yes, she can go anywhere." With this answer, the patient ventured out after a time. A similar renewed insensibility and fall ensued. The writer was summoned, and, having learned from experience that such fits of coma were sometimes caused by uterine irritability, proceeded to inquire ; and found pain in head, back, and thighs, difficult defecation, no dysmenorrhœa, and some degree of numbness of the extremities at times. The eye was clear, the head cool, the pulse normal, the complexion did not indicate cerebral lesion. A physical examination revealed a complete retroversion. Upon treatment with the

loop pessary, the headache disappeared ; the general
health improved, there was no return of the seizures, and the patient recovered.   Now, which physician pursued the wisest course ?   The first was
an active, intelligent, acute practitioner ; but his
preliminary education included no special gynæcological study.   Could he have been expected to
do differently ?   Probably thousands of physicians
would have done the same thing.   Had it not been
for special advantages, the writer might have done
worse.

A case was lately related where a lady had
been laid aside a long time with supposed disease of the heart.   She was treated by an intelligent, regularly educated physician.   Another
physician discovered a uterine lesion, which, being treated, the patient recovered.

Yet another case was that of a young married
woman, mother of two children, the youngest
about six months of age.   She suckled her infant, who throve finely, being a beautiful exhibi-
tion of infantile good health.   She complained
of pain in the top of the head, and in the back
(lumbar region).   It hurt her to lift her heavy
infant, troubled her to walk, and she felt very

miserable. She consulted the physician who attended her during her confinement. He made no vaginal examination, but gave medicine. There was no improvement. When her case was submitted to the writer, the uterus was found strongly retroverted. The organ was replaced by means of the uterine sound, and the length of the posterior vaginál wall was measured by the vaginal sound for the purpose of ascertaining the proper size of a pessary. *Upon a subsequent visit to apply the instrument, the uterus was found to be nearly* in situ naturali. *The pessary was worn only a few days, and the troublesome symptoms have disappeared.* This case shows the value not only of proper diagnosis, but of an *early recognition and treatment of a uterine displacement,* — a topic to be insisted upon farther on.

- The writer is glad to notice an increased attention to this subject on the part of the profession, thanks to the labors of Thomas and others ; but if it were general, and women were instructed, and understood the value of an early and prompt recognition of uterine displacements, — if they sought and received relief *early,* — what an immense amount of untold misery and suffering would be averted, and how much happiness would be introduced into families !

Additional Remarks upon Diagnosis. — Vaginal examinations being so absolutely essential to the full and clear understanding of the subject, a few remarks may be excused, in addition to what has already been said.

Position of Patient. — This is not a trivial matter.   An ordinary lounge or bed may answer for a base of position ; but an object that presents a softish, yet inflexible and smooth, surface is decidedly the most preferable.   A dining table, covered with a folded "comforter," answers well. Two dry sinks placed back to back will afford a similar desirable support.   The thighs are not buried in the support, as in case of the lounge or bed (which are also too low down for comfort). The height allows also of a vertical position of the examiner's body, without the unpleasant bending which congests the head when an ordinary bed is used.   The writer has found his chair an excellent medium for vaginal examinations.   The leg part is dropped to a right angle with the thigh portion, which is horizontal.   The back is turned slightly upward ; the patient then reclines on her side, and is turned up by the crank to any desired angle with ease and comfort. In this position the light penetrates more readily than when a horizontal support is employed.

Where a table cannot be provided without too much trouble, it is a good plan to introduce a piece of board under the upper mattress; an ordinary cutting-board of the sewing-room will answer well. It is very advantageous to divest the patient of her tight clothing, — corsets, hoops, drawers, &c. The writer remembers a case in which he was foiled in replacing a retroversion, until the under-clothes were the only ones worn. When the abdomen is artificially confined above and in front, where can the floating viscera go when attempts are made to elevate a uterus upon which they are so closely packed? The patient should lie upon her left side, head low, left arm protruding behind, knees drawn up close to the abdomen. The left side is advantageous, as it allows of full play of the operator's right hand. If he is left-handed or ambidextrous, the position on the right will answer. If the dorsal position is employed, the buttocks should project over and beyond the support, as there must be room for the sound to sweep about. The lateral position is less shocking to the patient's feelings.

The explorer should be provided with a uterine sound, *soap*, warm water, and towels. Lard, olive oil, or glycerine, may be used as lubricants. This is not a matter of trivial importance, as I once lost the care of a patient by not using glycerine, which she preferred. Soap is preferable, because it is so easily washed off. When lard is used, soap must also be employed to cleanse the hands. It is a saving of time, then, not to use the lard. If the vaginal secretions are copious, they will suffice sometimes for lubrication. If leeches are to be used, better employ water freely. Taking the natal furrow as a guide, the finger may glide over the anus, noting as it passes its condition as to fissures, hemorrhoids, prolapsus, etc. Or the finger may approach over the pubis to the vulva, noting the condition of the labia, clitoris, vestibulum, and urethra. It is then passed into the vagina, and swept about with care. The facts thus revealed are of great importance, and are as follows : *Condition of vagina*, — vaginismus, ulceration, irritability, tenderness, dislocation of axes, distortion of form, abolition of normal functions. The

abnormal condition of the axes is almost invariably
present in any case of version, flexion, or procidentia.
I say *almost*, because I have heard, but never have
seen one, of such cases. The finger next notes the
existence, position, size, and condition of the os
uteri. The os is the orient point of diagnosis in
malposition. It may be obliterated so as to defy
discovery. The uterus may be absent. When
existing, the os may be high up, low down, in the
concavity of the sacrum, over against the pubis, on
the right or left side of the vagina, in the vulval
outlet, or outside. The existence and thickness of
the lips of the os, and any solution of continuity,
excrescence, abrasion, hardening, or patency may
be discovered by the finger, and confirmed by the
speculum examination. Having thus made a defi-
nite exploration of the os, the digit should follow
the cervix anteriorly, posteriorly, and laterally. If
the os is in the uppermost part of the vagina, in
the central axis of the pelvis, with *cul-de-sac* well
developed, and no unnatural hardness on or around
it (a touch of the tip of the nose gives the best
idea of the feel of a normal uterus), one may infer
a normal position and condition of the uterus and
perimetritic neighborhood. If there is a continuous

cervical hardness (or softness), a smooth bulging, an occupation of the upper part of the vagina, the os somewhere in the concavity of the sacrum, it is fair to infer *anteversion.* The diagnosis may be confirmed by the introduction of the uterine sound, and the behavior of the parts.

If there is a similar condition, and the os points towards, or under, the pubis, *retroversion* is to be inferred, — subject to sound confirmation.

If the finger finds that there is an angle formed by the neck and body of the womb, — always a rentrant angle, never a solid, pointed one, — os toward sacrum, we have *ante-flexion.* If there is the same condition, os pointing toward pubis, we have *retro-flexion.*

If there is a turning of the fundus to one side, we have *latero-version.*

If a bending, *latero-flexion.*

Indurations, which are liable to be mistaken for the uterine substance, may be caused by pelvic *areolitis* (often erroneously termed *cellulitis*), intramural tumors, prolapsed ovary, ovarian tumors, rectal accumulation or enlargements, pelvic hematocele, cysts of the broad ligament, piles, prolapse of vagina, malignant disease, hernial protrusion in the vagina, etc. But the careful use of the uterine sound will distinguish them generally, although sometimes it is necessary to dilate the uterine cavity with tents. The use of the uterine sound is fallacious sometimes, as will be shown farther on. But when a flexed uterus is replaced by the sound, and held there, a digit passed about the cervix will easily distinguish its normality as a general rule. It will feel clear, well-defined, distinct, and mobile. The concavity of the *cul-de-sac* will have a softness and elasticity which are absent when there is perimetritic disease. All the procedures named should be made with gentleness; coaxing the tissues answers better than force. The amount of pain produced is a very good general guide as to the amount of metritic or vaginal inflammation. Hence, the examination had better be conducted without anæsthetics, if possible. Vaginismus and irritability would disappear under partial etherization, while inflammatory pain would require more full etherization.

Circumstances under which the Use of the Uterine Sound has proved Fallacious. — So much reliance in this treatise is placed upon the use of the uterine sound as indispensable in the treatment indicated, that it is proper to allude to the fact that the evidence afforded by the use of the uterine sound is not *always reliable.* It is liable to misinterpret conditions, just as the senses are to misinterpret objects. This is no rational argument against its employment, any more than that the eyes should be disused because of themselves they are not able to give us an accurate idea of distance and shape. However, as it has been found to be mortifying, if not dangerous, for physicians to have made wrong diagnoses by the use of the sound, thus exposing themselves to actions in court for malpractice, it is thought best here to call especial attention to the fact that a uterus may be enlarged by pregnancy, or hypertrophy, or diseased conditions, *and yet the uterine sound faithfully introduced gives no idea of the presence of an ovum, or ova, or diseased condition in that organ.* The most eminent practitioners have been thus misled when they really felt they were making a careful examination.

Cases. — A middle-aged married woman had a uterine fibroid of considerable size, involving the posterior uterine wall, and protruding into the vagina behind. Her menses had paused. She had considerable malaise, and distress, and nausea; but inasmuch as she had had the same symptoms before, it was a question whether she was pregnant, and, if so, whether abortion would not be justifiable. Two physicians were consulted. A sound was carefully passed into the uterus. It took a normal course, passed without friction, and was arrested at the depth of two and one half inches. Both physicians confirmed this measurement, and decided that there *was no pregnancy.* The issue showed that there was an ovum in this womb, for the woman miscarried within two weeks, — thus settling both questions adversely to the decision of her advisers. This case happened to one of the most eminent physicians in Massachusetts.

A second instance of misinterpretation by the uterine sound happened in the case of a young woman nearly dead with consumption. There was so much distress referable to the pelvis that the vagina was explored. A hardish substance (continuous with the os which pointed backward) was found forming the vaginal dome. The uterine sound freely passed into the cavity of the womb by carrying its handle back beyond the body. It was arrested so that only two and a

half inches were buried inside the organ. No
further procedures were instituted on account of
the low condition of the patient. After death,
the uterus was found to be about four inches in
depth. The walls were thickened, and its cavity
was occupied with a whitish semi-solid cheesey
mass, which was readily broken up by the fingers.
On the top of and by the side of the uterus was
a dermoid ovarian cyst, globular in shape, parietes nearly white and solid, containing a fluid of a
golden straw-color, a collection of hair, and a cit-
ron-colored solid mass the size of a goose egg, — a
state of things wholly unexpected, and contrary to
the given diagnosis, — *anteversion!*

Another instance is furnished by the case of a
young married woman who had anteversio-flexion
with vaginismus, and enlarged and tender uterus.
This was relieved by treatment, and the patient
not seen only at odd intervals. She had a renewal
of the old symptoms subsequently, and the attention
of the physician was called to the case again. A
physical exploration of the vagina revealed the os
towards the sacrum, a vaginal dome continuous with
the os, and of similar feel with a rentrant angle
well marked. The vaginal sound passed readily

into the organ, with some suffering, revealing the
uterine cavity three and one half inches in length.
The organ was easily elevated into position. It
was tender on handling. As before, leeches were
applied; the symptoms continuing, counsel was
called in eleven days afterwards. The sound was
then used by the consulting physician, and local
depletion was adopted by puncture, which was repeated several times afterwards. The question was
raised at the consultation whether pregnancy existed. The evidence afforded by the sound was regarded as sufficient evidence to the contrary, as no
such penetration, it was supposed, could be possible
with an impregnated uterus. Seventeen days after,
twin fœtuses were expelled! Thus the fallacious-
ness of the physical signs was exposed, resulting in
the institution of a lawsuit for damages by malpractice against the attending physician; who will probably never again need to be taught that the uterine
sound, though apparently simple and truthful, cannot
always be depended upon in estimating either the
size or contents of an enlarged uterus. There are
circumstances under which the usual signs furnished
by auscultation and percussion prove fallacious.*
The department of uterine physical exploration

* *Vide* Boylston Prize Essay, 1857.

must be added to that of the thoracic, as furnishing instances of fallacious reasoning. The physician should not be dogmatic or too confident in any department of physical exploration, as it is human to err.

An instance has been reported where the sound has penetrated the walls of the womb, and passed out into the abdominal cavity, no harm resulting! It is probable that the very careful introduction of the sound into an impregnated womb allows of a separation of the membranes from the inside of the uterine wall, and thus permits the sound to traverse it until it pockets in the uterine wall, and gives the impression that the fundus is reached. There may be a way of obviating this, but it is unknown to the writer. In cases of enlarged, flabby, verted uteri, whose lower wall is bedded in or attached to the surrounding tissues, it is easy to conceive how a fallacious replacement may be made by the sound carrying before itself the upper free wall, and stretching it so that the uterus will appear to be in place, because the sound occupies the normal direction. But this fallacy may be detected by the non or partial rising of the lower wall of the uterus, watching by digital touch as

the elevation is practised. This subject of falla-
ciousness of physical signs is one of cardinal importance in the treatment of versions and flexions
by mechanical means, as such means will be sure to
fail and come into disrepute if complicated with
pregnancy or diseased conditions, — metritic, inter-
metritic, or perimetritic, — unless such conditions
be specially recognized and appreciated. Indeed,
much of the present disrepute of pessaries comes
from this very circumstance ; and hence it is that
the writer respectfully requests that no one employ
his pessaries without consulting this little treatise,
and obtaining a clear idea of what they are intended
to accomplish, and what not, by their inventor. In
this way the best results may be brought about,
and perhaps some improved mechanical method of
carrying out these ideas may be suggested to the
mind of the reader by some power from without
(ideas, as well as native faculties, are gifts), as the
perusal may awaken new trains of thought.

Ætiology. — This department of diseased actions is so large, that volumes could be written
upon it alone. Indeed, it is so very comprehensive and varied, that one always approaches causes
with reluctance, for fear of not getting hold of the

right ones, and thus give erroneous impressions.
It is just so with other physical events, however
simple and plain. For instance, a cannon is fired.
The question is, Who fired the cannon? The
commander says, "I fired it, because I gave the
order." The gunner says he fired it, because he
pulled the string of the percussion-cap. But the
men who loaded the cannon, those who made the
gun, powder, cartridges, dug the ore from the earth,
— in fine, any one who had any thing to do essential to the existence of the piece, — might truth-
fully claim the right of having fired the gun ; as,
without any one of them, this particular piece of
ordnance would never have been exploded at the
given time and place. It is just so with disease.
The fibres and threads of its warp and woof run
immense distances, for long times, and are inex-
tricably mixed up. They all have to do with the
impaired function and changed physical condition.
Still, for all this, there are a few causes so positively marked and generally acting, that, although
somewhat uncalled for in this special work, — because so fully and clearly dealt with in the systemic works on diseases of women, — attention
is asked to them because of their relation to

the well-working of the instruments here brought forward.

Dress. — There is no doubt in my own mind that the present mode of suspending the dress of females from the waist, and the use of corsets, is a most prominent existing cause of uterine versions and flexions. The best authorities coincide in this opinion. It cannot be too strongly insisted upon, that the natural points of support for suspending the garments in men and women lie between the tips of the shoulders and the neck. The bones of this region, with their investments, are admirably suited for this purpose. The clavicles, the scapulæ, and the ribs form a natural framework for support ; weight applied here is supported by the whole thoracic and pelvic skeleton. It is borne without at all interfering with the contents of the great cavities of the body, because it is so evenly distributed over a large surface. There is no interference with the diaphragm or abdominal viscera. It is, indeed, the *only proper way* to suspend clothing. On the other hand, waist suspension, combined generally with corsets, no matter how loosely worn, none the less throws the superincumbent weight of the garments on to the abdominal region, and crowds the viscera down to the lower part of the cavity in the pelvis, for the simple reason that the viscera can go nowhere else! They cannot go upwards, the diaphragm prevents this; they cannot go backwards because of the spinal column, lumbar muscles, and ·short ribs; they cannot go forwards, the naturally yielding abdominal muscular walls are kept in by the corset and waistbands. If they move at all, they must crowd *downwards* into the pelvis, because it is the lowest part of the abdomen and the weakest in the female. In this state of things, let the vagina be weakened by inflammation, what would be more natural than for the uterus, unduly weighed down, to tip over or bend, thus dilating still more the toneless vagina, and increasing the difficulty? See how disastrous this waist suspension is in men, although their pelvic cavity is smaller and stronger than in women. Take the seamen. They are notoriously subject to hernial protrusions. No doubt, their unusual efforts in pulling ropes combine to aid this result; but the tight waist-belt must make it sure. The abdomen is a plenum, and subject to laws of hydrostatics. Pressure is distributed all over it, and, if there is

any giving way, it must be in the weak spots.
The pelvis of the female is the weak spot; why,
then, persist in hanging clothes where it is con-
fessedly injurious? When Paris fell, it was hoped
that the fashions would go with it, and that the
common-sense of mankind would cause them to
look for modes of dress from medical artists, who
understand the needs and requirements of the body
from a physiological as well as æsthetic point of
view. Health and strength should be combined
with true beauty. The person who will invent and
successfully introduce a means of suspending the
garments of women from the shoulders, which shall
combine ease, lightness, and mechanical adaptation,
will deserve the reward due to a public benefactor.
For these reasons, changes from any quarter are
welcome, and it is a pleasure to note that great
progress is now making in dress-reform, which, it
is understood, embodies the principle of shoulder-
suspension. Dr. Thomas recommends and pub-
lishes a cut of Bacheller's skirt-supporter, with a
circular piece of a thin band of metal surrounding
the waist. Probably it can be obtained of any instrument or bandage-maker.

The idea is to have the clothes suspended from

the shoulders, as the pants of little boys are from their waist-jacket. It is best done by buttoning the drawers and skirts on to a jacket fitting the bust. To preserve the bust, a small girdle may encircle the bosom.

I can recommend the practical results found at the Dress-Reform rooms, 24 Winter Street, Boston, from a trial in my own family. It is a move in the right direction.

Impoverished Food as a Cause or Promoter of Uterine Disease.—Vegetables and animals suffer by being fed upon impoverished aliment. If plants do not receive the proper amount of manures (which are not much more than vehicles for carrying mineral elements to the plant), as every one knows, there is a feeble, fruitless vitality. A dog fed exclusively on flour by Magnedie died in forty days, and dogs died in the same time with no food at all! Dogs fed on sugar, exclusively, lived about as long as those which were fed on flour. The sugar-fed dogs had ulceration of the corneas, followed by an evacuation of the humors of the eyes and blindness. So that the development, tonicity, and existence of the tissues of the animal body (other things being equal) depend upon the food.

The training of pugilists and boat-racers shows
this remarkably. There is a wonderful improvement in muscle, mind, and mental elasticity. On
the other hand, the effect of the withdrawal of
food is shown in the history of the rebel prisons in
our late war. It is impossible to conceive of a perfectly healthy human body without a perfect nervous system, producing force, neurotic power, or
whatever you may be pleased to call that influence
which is generated in the ganglionic nerve centres,
and is distributed throughout the body by the
nerves themselves and their terminations, especially
for the purpose of "running" the physiological
machine, whether it be by repairing waste tissues,
or building up new ones. Without nerve force the
body is a dead machine, like a locomotive without
steam. In a certain sense, nerve force digests, as-
similates, and appropriates the food, and presides
over (besides the functions of cerebration) phona-
tion, menstruation, reproduction (in all its phases),
&c. If a woman's nervous system is perfect,
and perfectly nourished, she probably will have
a *perfect* physical condition of the uterus, as this
is *a* if not *the* most important nerve centre of
her system after puberty ; or, if this proposition be

denied, she may bear up against, or carry along
without much trouble, deviations of the uterus, —
because her perfect nervous system is strong enough
to resist its drawbacks, — or the injudicious, accidental, or physiological efforts that vert and flex the
uterus before, during, and after their action. Take
a nation of women who have their nervous systems
perfectly nourished, — would it be natural to find
them with deviated wombs; or, if they had them,
would they not bear up under them, and not show
signs of malaise? No doubt, versions and flexions
have existed ever since the race was created ; but
why should they be so prevalent of late years?
Making all due allowance for imperfect diagnosis,
is it at once to be supposed that the acute and
grand minds of the past generation of physicians
would have overlooked them, if so common then as
now? An intelligent English nurse, in one of the
smaller but perfectly equipped hospitals in Paris,
said (1862) that her observation, while previously
engaged in practising her profession in England,
taught her that uterine diseases were terribly
prevalent there, and that in France she found it
just the same. She inquired as to American
women, adding that she hoped they, at least, were

more free from these diseases than their English
or French sisters. When obliged to receive the
answer that there was no immunity of prevalence
here, she asked in despair, " What is it that ails
women, and causes them so generally to be afflicted
with such diseases ? " Not being able to give an
answer to this nurse's query, it has rung in my
ears more or less ever since, without any thing like
a satisfactory answer until lately. The fact that
flour is impoverished food, and that it is so generally used as an article of diet by civilized nations
more than any other one article, and that this
prevalence of flour-eating coincides with the prevalence of uterine diseases, — this indicates that
the two may be connected together, as cause
and effect. Contemporaneousness and sequence
are not always causation. Still, effects do follow
causes. If a woman with weakened vagina and
uterine ligaments practises tight lacing, hangs
her clothes around her waist, and is afterwards
found with displaced womb, we see that mechanical forces have been at work, and can clearly
trace the connection between a constricted and
weighted abdomen and the " squashed " uterus.
And, if on inquiry into the causes that weakened

the vagina and ligaments, we find the female in question mainly lived upon a food which the chemist* tells us contains only one-quarter of the ash of wheat, $\frac{21}{82}$ of the phosphoric acid, $\frac{1}{6}$ each of the soda and lime, $\frac{5}{22}$ of the magnesia, and no silica, sulphur, or sulphuric acid, which are found in wheat, — a general impoverishment of mineral ingredients of $\frac{3}{4}$, — may we not reasonably infer, if we cannot conclusively affirm, that so exclusive a consumption of weakened food was no insignificant cause of her vaginal and uterine infirmity ; and hence, in general, that prevalent *flour-eating* is causatively related to prevalent uterine diseases ?

We wish to be understood not as asking whether this impoverishment is *the* cause, but as *one* cause, in the production of uterine diseases. We know it does have some effect upon the mammary glands, in diminishing the quantity of their secretions. We are acquainted with one female of middle-age, with her ninth child, who has a very abundant supply of milk, upon which her novene offspring has thriven so finely as to possess a symmetrically developed system (the head preserving the proportion with the trunk that is seen in adult life). Before and

* Johnson's "How the Crops Grow," New York : 1874.

since the conception of this child, the mother eschewed flour in her diet. With the exception of the first two children, this mother had a deficiency in the mammary secretions. During this period she lived upon flour. When investigation is made into the methods practised by dairymen to increase or maintain the quantity (not to say quality) of the milkings, it will be found that the kine are fed largely upon "shorts," or "fine-feed," or bran, — called in England "pollard" and "sharps." That is, the animals are largely supplied with the tegumentary portions of wheat, which are rejected in making flour, and which contain a large amount of mineral ingredients necessary for the production of milk by the epithelium of the mammary gland. When it is considered that milk is the only article of diet that combines solid and liquid human food in one substance; that it contains also all the mineral salts that are found in the human body (in this respect unlike any other glandular secretion), would it not be natural to expect to find diminished lacteal secretion when the food of the producer was deficient in those salts which are normally found in milk?

Among other cases, an instance of copious lactation under unfavorable circumstances (except the wheat diet) is furnished by a married lady, thirty-four years of age, — pronounced by Dr. Bowditch to be in consumption at the age of twenty-four. Her first child was born a few months ago (Nov. 1875). She was advised to wean this child, soon after birth, as, owing to the affection of her lungs, it was asserted, " You will have to do it soon, and you had better anticipate the inevitable." The reply of the mother was rather indignant, as she said, " Do you think I shall wean my child, when I have so much milk as to be obliged to throw it away?" This lady lived entirely upon animal food, supplemented by the cookery made from "whole wheat meal."

When we hear that Indian women suffer, like their civilized sisters, with prolapsi, versions, and flexions of the uterus (testimony of Surgeon J. J. Milham, U. S. A.), we are forced to conclude that their apparent immunity from the disabilities that follow such lesions in civilized society is due to their mode of life.

We would not be dogmatic on this subject, — only suggestive. The vitality and perfection of plant structures depend largely upon the proper abundance of their mineral food as furnished by fertilizers, manures, virgin (that is, unimpoverished) soils, &c. If you would have potatoes of feeble vitality, disposed to scarcity, wateriness, and early decay, you have only to allow them one-quarter of the nourishment ordinarily deemed necessary. A similar treatment of a human constitution will produce corresponding results.

We therefore feel justified in affirming that the necessity of a good general diet cannot be too strongly enforced or insisted upon to preserve female health, or to restore it when impaired.

Although this is an almost self-evident proposition, we recur to it because proper diet forms the basis of the treatment of chronic disease. It is the foundation on which to lay the superstructure of good health. It is almost impossible to have the tissues of the body healthy, unless they are supplied with a sufficiency of food which contains *all* the elements, or ingredients, that are found in these tissues. This is an axiom in physiology. It also is an axiom of common-sense. No nerve, muscle, bone, sinew, vessel, or organ can thrive, unless fed with the elements that enter into their composition severally. And, when diseased, medicines cannot supply these interstitial elements where 'deficient. The medicines may create an appetite, and thus prepare the demand for food ; but it is the food that restores, corroborates, strengthens. Food is fuel. Food is force, — the potential force of the sun, ready to become actual energy in the manifestation of vital functions ; among which the *restorative function* is one of the most impo†tant in disease.

What, then, is *good diet?* Most will reply, *starchy foods*, — corn starch, or maizena, tapioca, sago, flour-gruel, arrow root, &c. These are universally regarded as " light " food, and hence more easily digestible. They also have the advantage in an æsthetic point of view. They *look* white, and nice, and delicate ; and hence are naturally deemed of more dietetic value. Why *looking white and nice* is any standard of nutritive excellence, is more than can be explained. It would seem that the standard of nutritive excellence should be based upon some such features as follows :—

(r) Assimilability.

(2) Nourishment, — the furnishing such elements as are contained in and needed by the tissues of the body.

Assimilability and nourishment are, then, the

characteristics of good diet. The question of "lightness" and color are really questions of æsthetics, not of dietetics.

(1) *Does Starch stand the test of Assimilability?*

According to the best authorities in physiology, starch must be changed into sugar — from a colloid into a crystalloid — before it is capable of being absorbed. This conversion happens not in the stomach, but in the intestines. It is a slow process. It takes time and chemical changes to effect an assimilation of starch, as compared with nitrogenous or animal food which is mostly digested and assimilated in the stomach. Suppose the powers of digestion are impaired, starch, after changing into sugar, undergoes a process of fermentation ; carbonic acid gas is largely formed, — called "bloat," or "wind" in the bowels, — which regurgitates into the stomach, distending the same, and deranging both. We are indebted to Dr. Salisbury, of Cleveland, Ohio, for the suggestion that the carbonic acid gas acts as a local anæsthetic upon the villi and muscles of the intestines, paralyzing them, and thus arresting the processes of intestinal absorption, and still further

distending the guts, because the muscular fibres
are incapable of contraction from this very paralysis. He also accounts for the cardiac disturbances, by an absorption of the carbonic acid in
excess, and acting upon the very cardiac muscular
fibre itself! Several years' experience enables the
writer to confirm this opinion of Dr. Salisbury.
Under these circumstances, the ingestion of starch
seems a very hazardous thing, by keeping up the
very fermentation which is the cause of so much
trouble.

(2) *How does Starch stand the test of Nourishment?*

The chemist tells us that starch is isomeric
with sugar. Its formula is $C_{10} H_{12} O_{12}$, — three
elements; but the body tissues need some fifteen,
at least. Here, then, are three elements to make
up nourishment where *fifteen* are required! You
can make fat with starch, as the formula of fat is
about the same. But is fat a healthy constituent
alone? In excess it constitutes disease, as in
apoplexy, heart-thinning and weakening, some
forms of Bright's disease, arcus senilis, &c.
Women with uterine disease are very apt to look
fat, and therefore some hastily conclude they are

not sick. It is possible to get fat too easily. Starch taken alone must be *very poor food.*

Poor, — because deficient in twelve ingredients that should enter into normal food.

Poor, — because it takes so long to digest, and after it is digested can only make fat, which becomes a disease when formed in excess.

Poor, — because it so easily ferments and forms carbonic acid in excess, to add to the plagues of the unfortunate patient.*

From these simple statements, it appears that the popular opinion of starch as good diet for the sick is not founded upon the truth ; and that the white color and good looks — æsthetic excellences — are not dietetic excellences.

It is recommended to use *animal food,* such as milk, eggs, fish, flesh, fowl, &c., because, —

(1.) They contain all the fifteen elements at least that are necessary to sustain the body ; and

(2) They are digested mainly in the stomach.

When food is put there, the liquid portion is ab-

---

* Ex-Surgeon-General W. A. Hammond once tried to live upon starch alone ; but he so rapidly failed, that his friends interfered and induced him to stop before disaster was insured. — *Prize Essay, American Medical Association.*

sorbed directly. Then a large quantity of gastric fluid is secreted from the blood and thrown into the stomach. This soaks food and dissolves a portion of it. When all is dissolved that can be, the charged fluid is then absorbed into the blood, and carried through the portal circulation. After this process is complete, a fresh amount of gastric fluid is poured forth, soaking (digesting) the food, dissolving another portion of the animal food, and being then reabsorbed. And thus this collateral circulation is kept up until all that the gastric juice will dissolve is made soluble, the remainder passing into the intestines, to be acted upon by their juices.

In the treatment of uterine tips and turns, we advise that a diet of animal food should predominate. Let starchy food take a secondary place. The chances of success will be greatly enhanced as in no other way. At the risk of being tedious, this is insisted upon. The uterine neuroses can be best borne with a diet of animal food.

Reference is here invited to the Mixed Diet, (*vide* page 42), as one that has been found practically useful in uterine diseases.

The fact, that in three cases of uterine fibroids a diminution occurred under the mixed diet deserves attention. In one of these, — a flour-eater, — the fibroid extended beyond the umbilicus, and receded more than one-half in ten months. More experiments in this direction are needed. It is certain that the uterus can be affected by food.

Over-work. — It is possible to conceive of women maintaining their bodily health under unfavorable circumstances, if they are not *over-worked.* Work is the expenditure of force to overcome resistance through any space. Life is made up of such expenditure of forces. It is evident that the amount of force inherent in any person is limited in amount. According to late physiologists, the grand doctrine of the conservation of forces has been applied (or rather ascertained to act) in physiology. By them it is stated that force (confined to the human system) is generated from the potential energy of the sun, which enables the plant to form, directly or indirectly, articles of food, by decomposing the carbonic acid of the atmosphere, appropriating the carbon in the forms of food, and setting free oxygen. Now when we take into our systems direct or indirect vegetable food (carbons), and at the same time inspire oxygen, the force evolved by the combination of the carbon of the

food and the oxygen of respiration forms the
actual or kinetic energy to be used, as systemic
work in order to develop and maintain the body
and perform its functions. During work, carbonic
acid is given off into the atmosphere, to be taken up
again by the plant, by the sun's influence, appropriated as before described, — and the circle of changes
is thus complete. The forces being derived from
our food, it is evident that the limit of work-power,
other things being equal, is fixed by the amount of
food appropriated and assimilated : hence it is scien-
tifically possible to conceive why there is such an
over-work, or using up more force than is furnished
by the food. If this procedure is kept up, there
naturally results a rapid diminution of vital force.
The stomach and intestines suffer with the other
organs, and thus are rendered unable fully to
digest the food which may be put into them. Fermentation ensues, followed by the production in
large quantities of carbonic acid gas. When
absorbed through the stomach, diaphragm, and
pericardium, it produces a paralyzing effect (local
anæsthesia) upon the intestines, the stomach, and
the heart itself (Salisbury). The diseases which
this chemical process causes are well known, — as

colics and palpitation of the heart. This gastro-intestinal paralysis is of immense detriment to the already over-worked patient. Or, to take another view, the vital force required to. digest food under such circumstances must be increased, — it is such hard work as compared with that required in health. So that the results of over-work from deranging the nutrition of the body are seen in a weakening of the tissues of the body, of which the genital organs are an important and easily-affected part. Constipation of the bowels has been relieved by elevating an anteverted uterus. A lady who had had gastric trouble, so that she could not eat a meal of victuals without distress for eight years, was immediately relieved by the elevation of her anteverted uterus. It is impossible to do justice to this part of the subject without transgressing limits. Attention is respectfully in-vited to a few of the various forms of over-work to which females are subject, and which render them liable to, if they do not cause, affections of the uterus : —

(*a*) Standing posture in schools or stores.

(*b*) Pregnancy combined with too much employ-ment.

(*c*) Going up and down stairs excessively.

(*d*) Non-observance of rest during menstruation.

(*a*) *Standing Posture.* — It is one of the blots upon our modern civilization, that so many of our cities contain shops where female clerks are employed, who are obliged to stand all day upon their feet with no interval of sitting down.

The case of the female school-teacher is somewhat better than that of the shop woman, as she has more control of the matter, and can sit down if she chooses; but how many such even are to-day laying the foundations for future suffering, by disregarding the symptoms of distress in back, or body, or limbs, and continuing to exercise their profession in ignorance of these warnings! A school-teacher with uterine lesion, who has been laid aside from this physical disability, presents one of the most difficult cases to manage.

(*b*) *Pregnancy combined with Over-work.* — The treatment of pregnant animals is much more lenient than that of women. No work is exacted of the mare for some time before foaling. Generally she is allowed a free pasture-range, because over-work will injure the foal. But hardly any one thinks thus of the pregnant woman in the family of the average citizen of to-day. She has no let-up in her daily toil, no vacation of leisure ; nothing but work, work, till the hour of labor, when her kinetic energies are engaged so thoroughly in the pains and throes of child-birth as that the word *labor* is used to describe this bodily effort in contra-distinction to all other work. No sooner does the puerperal woman recover from her ex-haustive efforts, than she is obliged to arise and perform difficult and heavy work, before the contraction of the uterine ligaments or the subin-volution of the womb have fully occurred ! Is it a wonder that such women are found with uterine displacements ? The good effects of care in such a case are seen in the writer's maternal grandmother, — a lady now ninety-two, remarkable for the preservation of her faculties, her industry, and her knowledge of the present age. She gave birth to nine children ; but in all her confinements remained within her chamber at rest, till the child was four weeks old.

(*c*) *Going up and down Stairs.* — It has been stated by an authority whose name is forgotten, that it requires *four times* the power to ascend stairs that it does to walk on level ground at the

same rate. Generally, stairs are ascended four
times as fast as progression on floors. So that the
amount (if our authority is correct) of work thrown
upon the heart and nervous system is increased six-
teen times at one bound upstairs! The effect on
the nerve centres is shown by the rapidly palpita-
ting heart, the short breath, and sweating sometimes.
And it is reasonable to expect, when a female runs
up and down stairs many times a day, that her nerve
centres must suffer, — most palpably the uterine
nerve centres, — because so much of her limited
force is thus exhausted. The exhaustion of the heart
is seen in men who perform similar pedestrian ex-
ercises. It is strange that, when there is so much
unoccupied land in the world, mankind should deem
it necessary to erect residences and stores of sometimes seven or eight stories! Savages have here
certainly the advantage over civilized races, for their
dwellings are of only one story. A true regard
for the hearts of men, and the uteri of women,
dictates an adoption of this custom from the
savage. — The authority alluded to also gave directions for ascending stairs, so that it could be done
safely. As trials of this rule have resulted successfully, and as no hopes are entertained of ever

seeing the present physiological iniquity of multiple-storied buildings abolished, it is here appended with thorough endorsement. It is founded upon the fact of the quadrupled increase of force necessary to ascend stairs : *"Between each step count slowly one, two, three, four !"*

(*d*) *Non-observance of Rest during the Menses.* — At this time the periodical congestion engorges the uterus, adding to its weight, and thus rendering it more liable to displacement ; and besides all this, the expenditure of nerve force in performing the function is a draft upon the energies of the menstrual woman, no matter what is said to the contrary. Under these circumstances, lifting weights is liable to occasion displacement. I recollect a young, single woman of fair proportions, and such abounding good health that she was regarded as a model of a perfectly healthy woman. She temporarily left our neighborhood, and on her return she was so changed that I hardly knew her. She was pale, moping, spiritless, complained of pain over the ovaries, in the back, and of general malaise. Her uterus was found to be anteflexed and very sensitive. Upon treatment, her health was restored. It appears that during her absence she nursed a brother's wife until death from consumption. She stated that she was obliged to lift the invalid often, and that during her menstrual epoch she made no difference in her lifting. The flexion was attributed to the lifting process, inasmuch as she had been in such perfect health before her duties as nurse began.*

Under-work.— If the evils of over-work are so great, on the other hand those of *under-work are dangerous.* Life is action, and rest, continued while the body is in health, is worse than laziness ; it is disease. The tissues under these circumstances undergo a fatty degeneration, and hence become weakened, impaired in function, and relaxed. This is shown in the fattening of animals. Fat is found everywhere lying between the muscular fibrillæ within the connective fibrous tissues, as seen in over-fattened ham. So that, to escape fatty degeneration and its consequences, persons must exercise in some form of moderate exertion suited to their strength and faculties.

A condition of perfect inactivity is thought to be the state of the blessed in the earth.

* See "Sex in Education," by Dr. E. H. Clarke, for more on this subject.

Never was a greater mistake. Spirited horses, of magnificent build, have been entirely ruined by not being used and kept well-fed. Their life and energy were lost by non-use. By exercise the carbon is burned off, and not left to accumulate itself in the form of fat infiltrating the tissue.

Moderate labor, which produces some good to the world, is heartily advocated for females, as the best preservative of their tissues and strength. Under-work is as injurious as over-work.

Importance of Early Attention to Uterine Disease. — A young lady lately complained that every step she took was painfully felt in her bowels, over the pubis. She had frequent micturition, some headache, but no leucorrhœa or difficult defecation. Her employment — music-teacher in the public-schools — caused her to be much upon her feet. She willingly submitted to a digital examination of the vagina, which revealed an anteversion with some prolapsus. The uterus was mobile, normal in feel and resistance. There was nothing abnormal discovered in or about the vagina but the anteversion. The uterus was reinstated by the uterine sound with facility. She walked away from the office with no

hurt at all. The pain did not afterwards return.
Only a few subsequent passages of the sound —
more for diagnosis than any thing else — were
necessary to a perfect cure. In this case, the uterus
was restored before the ligaments and vagina were
displaced and stretched out of tone ; and hence
no mechanical means were necessary for support.
Another case was that of a primipara. When her
child was three months of age, she applied to her
physician with a complaint of great pain and distress in the lower part of her bowels ; frequent
micturition, difficulty in walking, and leucorrhœa.
The bowels were not tender ; the pulse regular ;
the skin normal ; appetite good. On examination,
anteversion was found. The uterus was reinstated
by the uterine sound ; a measure was taken, and
an anteversion loop pessary was fitted. The usual
instructions were given. Two or three days afterwards, on calling, I found that she had removed
the supporter, — not because it had hurt her, but
because she felt so well that she thought she did
not need it ! The uterus was found in its normal
position. It remained so until the next pregnancy,
which followed soon after. She got up well, and
has had no return of the version.

At a meeting of the Suffolk District Medical Society, held some time ago, the writer presented his pessaries for inspection. In the course of some remarks afterwards, Dr. D. H. Storer, LL.D., stated that he frequently met with recent uterine displacements in his practice, and that they uniformly yielded to a simple replacement by the employment of a uterine sound, without other mechanical treatment. So that it is fair to infer from these premises, *that the earlier the treatment, the less mechanical interference is necessary.*

The writer thinks that the importance of the knowledge conveyed in this last statement, to females, cannot be over-estimated, in view of the prevalence and intractability of uterine displacements. *Had these cases been taken early, there is every probability that they all would have been cured in a short time.* What one of us cannot remember some female member of our family, — our own mother, perhaps, — who has passed a life embittered by suffering caused by neglected uterine deviation? And what would not have been given to have such sufferers relieved otherwise than by death! And then when one thinks by how small an expenditure of skill the whole malady might have been averted at the first, he has a feeling of vexation that the required assistance was not afforded at the right time and place. Such disasters result, then, from ignorance, — *ignorance on the part of those who have the education of females under their charge.* It is the right, then, of every woman to know that to delay the treatment of uterine deviations is fraught with pain, vexation, trouble, and incurability in the future. How is such ignorance to be dissipated? There is no more legitimate way than that physicians should, as they have opportunity, warn parents and guardians of the necessity of *not overlooking* the signals of distress that Nature holds out as signs of serious trouble impending, and of consulting a physician thereupon. It then becomes this physician's duty to attend to the case; investigate it and relieve it, — not by *placebo*, nor by turning it off, nor by ignoring it, nor by underrating the value of the signs of distress, — but by carefully exploring — under ether, if necessary — the vagina and uterus. It is not required to subject every female to examination; but in these early cases the symptoms are apt to be frank and unmistakable, and if the physician ignores them he is guilty, culpably guilty, of an offence against himself, his profession, and his patient. In not attending promptly to these symptoms, he violates his duty, his professional oath, and humanity, and often inflicts irreparable mischief upon his patient.

No sentiments of false modesty should prevent the female from making her ailments known thus early. Prevention is always better than cure. The longer the tissues are twisted, bent, stretched, or displaced, the more difficult it is to restore them to normal condition. Medicine will not do it. Mechanical difficulties need mechanical treatment. As has been shown, experience demonstrates the truth of these statements. We can only fervently hope that they will be remembered and acted upon from a humanitarian point of view, and thus health be insured to the young and rising generation of mothers in regard to uterine deviation.

Politico-Economic Value of Female Health. — The material wealth of a State largely consists in human beings. The value of these human beings depends upon their moral, intellectual, and physical force, or their capacity for performing work of the threefold character indicated. Over one-half of the population is composed of females.

To females the remaining moiety owes its exist-ence. Females have given to the world its great men. All have been obliged to come into the world in the same humble way, and from the same source. Can we, then, ignore the value of women to a State? Suppose all the mothers of a nation are sickly, feeble, ill, and distressed ; will their de-scendants be healthy, strong, well, and comfortable? It is a remarkable fact that the fecundity of fami-lies has very much diminished of late years. This is shown by the decline in the number of relatives attending weddings, funerals, Thanksgiving and Christmas festivals. Has the prevalence of uterine diseases any thing to do with it? A nation that is childless will soon become extinct. And how can we do more for our country than by striving to un-derstand and successfully treat diseases peculiar to women ; thus rendering them hale (hale has the same root as whole), whole, and capable of bearing children with healthy (whole again) minds and bodies, " like plants grown up in their youth, or like corner-stones polished after the similitude of a palace " ? It may be thought that these words are out of place and ill-timed ; but can we over-estimate the value of women's services to a nation? History

shows the effect of bad mothers upon the destiny
of nations. Is it not greater to *produce* a President than to *be* a President ? Can we, then, with-
hold from the female sex any attention which we
can render, to make them more valuable in their
most important place in the nation ? For these
reasons, — poorly put and weakly arranged though
they may be, — physicians are implored to give
their attention to female health.

The late Dr. Hodge, of Philadelphia, gave as his
opinion, that no woman or child in health would
complain without cause. " Don't," he says, " call
a woman 'nervous,' simply because you are not
able to find out what is the matter." Hysteria is
a complaint commonly laughed at, and the poor
subject of it is held up to ridicule, if not contempt,
because "nothing ails her." But lately these convulsions have been directly traced to some version
or flexion of the uterus, showing a physical cause
for the neurotic disease. Is it pleasant to recall
cases of women treated with cold water from a
pitcher, when we ought to have relieved the verted
or flexed uterus ?

Were it not for the curse of politicians, it would
be very proper for the government to afford some

13

patronage to investigations to be made in the cause of uterine diseases. When election-day comes, the authorities then know the value of a voter. We wish they would remember election-day after it is passed, and institute inquiries as to the reason why as many voters are not born now as in the days of the brown-bread era of New England. If the nation needs rulers of good executive ability, legislators of good intellectual capacity, officers born to command, citizens of· staunch integrity, scholars of large accomplishments, men able to endure work in agriculture, commerce, art, and applied science, — then they must have their women looked after more closely, as the sources of all the laborers in its service. If a portion of the money spent in war were devoted to peaceful, systematic inquiries on the part of the medical profession, it would, we think, be more profitable for all concerned.

Additional Cases. — A married lady, forty years of age, had nocturnal palpitation of the heart, numbness of limbs, great nervous irritability, double and difficult vision, præcordial pain. There were no pains or bearing-down efforts in the back or abdomen. Dr. Derby, the oculist, said the eyes

were the healthiest he ever examined. There was
no acceleration of the pulse, the' cardiac sounds
were distinct and normal, and the nutrition was
good. No other organ being found at fault, the
vagina was explored. The uterus was retroverted
and not enlarged. Under treatment with the retroversion pessary, the eye and heart symptoms disappeared. She afterwards became pregnant, had a
difficult labor, with subsequently some return of
symptoms, accompanied by retroversion, which
were relieved by the same means, combined with
depletion.

A virgin, twenty-eight years of age, complained
of deafness in one ear. Upon removing plugs of
cerumen, the hearing was restored. She also complained of dizziness, and at times sudden loss of
consciousness, accompanied by falls of the whole
body on the floor or ground. There was pain
in the back, disturbance of the bowels, so
much so that suffering had become a second nature. On a digital examination of the vagina,
retroversion and retroflexion were found. It appeared from her statement, that, when twelve
years of age, while walking upon a paled fence, she
slipped, fell downwards, and one of the pickets

entered the vagina, thereby impaling her. Extrication was followed by profuse hemorrhage and other disturbances. After a few days in bed the symptoms subsided. It is probable that the mechanical injuries thus inflicted were the chief cause of the version and flexion. This lady was much relieved by the treatment laid down in the preceding pages. There has been no return of the epileptic-form seizures, and the nervous symptoms have very much abated.

Cardiac Disturbance increased by Uterine Disease. — A seamstress, about thirty-five years of age, complained of palpitation and præcordial pain. She had also pain in the back, inability to walk a great distance, great nervous prostration, and difficulty in carrying on her business from sheer weakness. There was an increased area of dulness on percussion over the præcordium, a bellows murmur strong and distinct, and a quickened pulse. The uterus was found to be retroflexed and retroverted. Treatment relieved, but did not cure, the symptoms. It is probable that, had the uterine lesion been detected and treated early, the result would have been better.

Case of a School-Teacher who could not walk on

Account of Retroversion. — This lady was examined by her family physician, and her disease discovered, but with no attempt at her relief. She was obliged to ride to school ; she could not stoop over to tie her shoes ; general health otherwise — except neuralgia — good. Tonics were administered, with no relief. The prospect was of a complete disability. She was the sister of a very eminent chemist, who placed her under local treatment. At first, only a short pessary could be worn, such was the " squashed " condition of the vagina ; then a longer, and afterwards a still longer. The result has been very satisfactory. The power of walking and tying shoes was restored. Although the uterus could not be elevated to its normal site, still the relief afforded was sufficient to elicit the remark that she " never was better in her life."

Cases showing the Relief from Leeching. — A lady with a verted and sensitive womb had a severe headache at the time when she was about to have leeches applied to the uterus to relieve the tenderness. After the application of one leech the headache was relieved. Nothing was said about the connection between the headache and the uterine tenderness, so that it was not an imagined relief, as might be supposed. Another case of anteversion about the same time had tenderness, with pain in back, limbs, and head. Two leeches applied to the uterus removed its sensitiveness and relieved the pains. Another similar case of long-standing, characterized by pain in right lower third of chest front and back, with pain over the ovaries, was relieved by two leeches, though weakened by the loss of blood. The great trouble in all these cases is that the uterus, being the weakest organ, suffers whenever the system takes "cold," which all are liable to. The same law obtains in physics. A vessel subjected to a strain will give way at the weakest point. Now, if there were any way whereby a still weaker spot than the womb could be secured without injury to the general system, then, when the system, as we say, "took cold," the shock of the "cold" would descend upon this spot and not touch the weakened womb; which, by being let alone, could then better regain its normal tone and resistance, and not suffer as before. It would render less necessary the application of depletion. This is the doctrine of the old-fashioned Setons, which are now very generally disused.

It is not a good plan to use them where people live upon impoverished food ; but when patients are put upon proper diet, more approaching the standard of old times, the action of setons in such cases as described has been satisfactory. A small silk thread or strand is sufficient. It is passed through a fold of skin, near the lower part of the abdomen ; that is, the superior spinous process of the ischium. The pain of this size of seton is very much less than the large size of lamp-wicking, but it is sufficient for counter irritation. The two ends are simply tied together like a rowel in a horse, not drawn through. A permanent blister by mustard or flies may be kept in the same spot with similar, but not so efficacious, advantage.

APPENDIX.

Case of Hemi-antero Uterine Congestion. — Dr. H. L. Hodge in his comprehensive work, "Diseases of Women," Philadelphia, 1860, p. 264, says: "It would certainly be very difficult for an educated pathologist to imagine that during life any congestion or inflammation could be confined to one half of an organ so vascular as the womb ; and perhaps it would be more difficult for an accurate and experienced anatomist to demonstrate upon a *post-mortem* examination an induration from inflammation affecting merely the anterior or posterior segment of the uterus." He had been alluding to the attempts of some to account for versions by the hypothesis that, by congestion, &c., of the anterior half, anteversion would be produced, and retroversion by a similar pathological condition of the posterior half. Taking the motto Dr. Hodge himself adopts in the work quoted, — "*nullius addictus jurare in*

verba magistri," — the writer would like to relate
a case of a young lady of seventeen. She was a
person of apparent good health and remarkable
beauty. Some three months before her death, she
had a mild typhoid fever. In about three weeks
from the seizure she became convalescent, and
apparently regained her full vigorous health. One
afternoon on returning from school she partook
very largely of artichokes, just as they grew in her
father's garden. Before night she was seized with
vomiting, purging, colic, and prostration, — all
which were violent and unyielding. Death followed the next morning, despite all the efforts to
save. Of course, the fatal issue was attributed to
the eating of the artichokes. A *post-mortem* examination failed to reveal any appearances sufficient to account for the death. There was, however, a livid, lead-colored hue to the anterior half
segment of the uterus. The colored portion was
elevated a little above the anterior portion, which
presented a perfectly normal appearance and feel.
The line of demarcation was clean, distinct, well-
marked, and was exactly in the median line dividing the antero-posterior diameter into exactly two
equal halves. A dissector's knife could not have

more equally divided the uterus into two antero-posterior halves. The truth of this statement may be relied upon. I agree with Dr. Hodge that it is impossible to conceive of such a pathological condition; but it required no educated, accurate, or experienced anatomist to see the physically changed condition of this young girl's uterus, — for the change was as distinct as if the anterior half vertically of the whole womb *had been painted black.* I do not propose to use this case as an argument in favor of version by weight, but it does seem to me sensible to allude to it as a possible cause of *flexion.* If the uterus is capable of being congested so that it is turned blacker than the erectile tissue of the turbinated bones of the nose in smelling a savor good or bad, — as this accidental death from artichokes demonstrated, — why, then, is it beyond the limits of possibility to suppose that this congestion may weaken the tissue of this side of the womb, so that the healthy side pulls it over towards itself from the want of proper antagonism ? Generally, the convex side of the flexion has been regarded as the faulty side. It seems to me that the concave side of the flexion is the one where the trouble is caused by its over-contracting against its other weakened half. The flexion, in the fingers of a glove, are on the contracting flexor side of the hand. As the flexion continues, the vessels of the concave side are compressed, — nutrition is impaired, and thinning results from the want of proper circulation. It is difficult to conceive of the congested side expanding so as to crowd over the uncongested half, and bend it; though not impossible, as seen in the unequal expansion of two thin pieces of board glued together in opposite direction of the grain.

According to Dr. Bixby, eighty per cent of infant females have anteflexed uteri. But at puberty anteflexion is abnormal. The frequent occurrence of flexion should spur the physician to every investigation as to its causes. It is to be hoped that no one will swear too much by one master, as truth is stranger than fiction or supposition.

Remarks upon the Vagina as orienting the Treatment. — This work not being controversial, it is not intended to enter upon any discussion; but it may be remarked that the views taken by the writer are not in accord with those of Hodge or Thomas. These gentlemen regard the uterus as supported by the, —

(1) Broad ligaments (two).

(2) Round ligaments (two).

(3) Utero vesical.

(4) Utero sacral.

They throw out of consideration the vagina, as of no value as a uterine support, chiefly because

It is a curved tube ;

It is distensible ;

It is thin ;

It is weak.

Hodge bases his principles of uterine support upon an intra-vaginal basis. With him, the distension of the vagina is a matter of no account whatever. It does not orient the treatment. In this the writer differs from him. He agrees as to the influence of the four ligamentous supports, but does not dismiss the vagina as of no value. Indeed, if one carefully reads the descriptions of both the authors named, he will notice that Hodge's pessaries are described as "levers," and their *fulcrum, the vagina.* Now as the fulcrum is the most important factor in the working of a lever, and is used to bear all the weight, — *i. e.* the womb, and all the power, — how does not, then, the *vagina become the support of the womb,* if it is the fulcrum on which the lever rests in elevating the womb? So that upon the doctor's own showing the *vagina* is the main element of power in his system of pessaries! Practically, he uses the vagina for a support, mainly by distending it laterally. This lateral *distension* has been commented upon in a former part of this work as acting to increase the difficulty. Scattergood's elastic intravaginal pessary is the best I have seen. This rests against the pubis, and elevates the uterus from the post-utero vaginal *cul-de-sac.* But as it is adapted only to retroversion it cannot claim general adoption, — indeed, the pressure it exerts often is too great for safety. We have thus seen how these intra-vaginal pessaries, except Scattergood's, depend for success upon the vaginal walls. Indeed, how otherwise can the uterus be elevated except by vaginal means? It is pretty certain that those *uterine* ligaments cannot be affected mechanically, and no one would be justified in cutting down and altering them. No other resource is then left but to restore the vagina to its normal state, in the hope that, by improved diet and artificial support, the abnormal condition of the ligaments may be restored to health. We can do no more, no less. The vagina is sometimes expanded so as to fill the whole pelvis (in labor). What becomes of the ligaments at the full term of pregnancy it is difficult to say, as they seem to be obliterated. In the physiological functions of childbirth, we see greater changes of ligaments and vagina (any one who has performed podalic version can testify to this) than in any case of version or flexion. And yet these parts resume their natural shape and position after labor by contraction and absorption. It has often, with reason, been recommended to females to bear children, in the hope that the parts would resume their natural position afterwards, by the mother taking care to lie abed for three or four weeks after labor. Sometimes this is effectual. May we not, then, be encouraged to hope that a measure of success may attend efforts to restore abnormal ligaments and vagina, because existing to a less degree than in the parturient female?

Bloat. — Females with uterine disease frequently complain of bloated bowels. By this they mean a soft or hard turgid condition of the bowels, which comes on suddenly and as suddenly disappears without any discharge of flatus from the mouth or rectum. There is something singular about this affection. The distension seems to be gaseous, but its sudden unaccountable disappearance is curious at least. I have been inclined to regard "bloat" as the result of defective innervation, — a weakness in the digestive organs, like dyspepsia, but unlike it in the quick absorption of the accumulated gas.

INDEX.

A.

C.

D.

N.

O.

P.

R.

S.

V.

W.

Cambridge: Press of John Wilson & Son.

A CATALOGUE

OF

MEDICAL WORKS

PUBLISHED BY

JAMES CAMPBELL,

Publisher and Bookseller,

18 TREMONT STREET, MUSEUM BUILDING (*directly opposite Messrs. Codman & Shurtleff's*), BOSTON, MASS.

A CONTRIBUTION TO THE TREATMENT OF

THE VERSIONS AND FLEXIONS OF THE UNIMPREGNATED UTERUS. By EPHRAIM CUTTER, M.D. Twenty-four Illustrations. Second edition, entirely rewritten. 16mo, cloth. $1.50.

"This is an excellent pamphlet on a difficult subject, enriched with many diagrams of the uterine organs, and the pessaries recommended by the author. We do not remember to have seen a clearer exposition of the subject in any work, and can heartily recommend this for perusal." — *The Medical Press and Circular, Edinburgh, Jan.* 31, 1872.

THEORY OF MEDICAL SCIENCE: The Doctrine

of an Inherent Power in Medicine a Fallacy. By WILLIAM R. DUNHAM, M.D. 16mo, cloth. $1.25.

Dr. Edward H. Clarke, formerly Professor of Materia Medica in Harvard University, writes the author: "I have read your book with much satisfaction. It is always pleasant to see one combat an error with energy and skill. You make your points with great distinctness, and support them with great ability. It was always my endeavor, in my lectures on Materia Medica, to insist as strongly as I could that drugs possessed in themselves no occult power, but that what is called their physiological action is the result of the reaction of the system upon the drug, as the latter passed through the former."

C. E. HOBBS'S BOTANICAL HAND-BOOK of Com-

mon Local English Botanical and Pharmacopœial Names, arranged in Alphabetical Order, of most of the Crude Vegetable Drugs, &c., in common Use: their Properties, Productions, and Uses, in an Abbreviated Form. 8vo, cloth. $3.50.

"Every druggist and druggist's assistant should have a copy."

FILTH DISEASES AND THEIR PREVENTION.

By JOHN SIMON, M.D., F.R.C.S., *Chief Medical Officer of the Privy Council and the Local Government Board of Great Britain.* First American edition. Printed under the direction of the State Board of Health of Massachusetts. 16mo, cloth. $1.00.

"The undersigned members of the Massachusetts State Board of Health would respectfully, but earnestly, urge upon all persons the careful perusal of the following masterly essay by Mr. Simon, Chief Medical Officer of the Privy Council and of the Local Government Board of England. If the practical suggestions made therein were acted on by all citizens, hundreds of lives now annually doomed to destruction would be saved, and the health and comfort of the people greatly increased."

HENRY I. BOWDITCH,	R. T. DAVIS,
RICHARD FROTHINGHAM,	DANIEL L. WEBSTER,
J. C. HOADLEY,	J. B. NEWHALL,
	W. L. RICHARDSON, Sec'y pro tem.

Members of the State Board of Health of Massachusetts.

A COURSE OF LECTURES ON PHYSIOLOGY,

delivered at the University of Strasbourg. By E. KÜSS, Professor of the Faculty of Medicine. Edited by Dr. MATHIAS DUVAL, Prosector to the Faculty of Medicine at Strasbourg. Translated from the revised edition by ROBERT AMORY, Lecturer on Physiology at the Maine Medical School. 1 vol., 12mo, cloth. Illustrated with 152 wood-cuts. Price $2.50.

"We consider the book to be the best of all the later text-books of human physiology that can be placed in the hands of the American student, and we cordially urge its adoption as a manual by those engaged in teaching this intricate and deeply interesting branch of science. Though concisely written, many of its topics are quite elaborately treated." — *American Journal of Medical Sciences.*

"When we say that, in our opinion, it is the best book that a student can procure, we only state that which is not only our firm conviction, but that of other teachers also." — *Medical Press and Circular.*

"M. Küss was one of those modest scientists who love science for itself; who seek for the truth without caring whether they are well spoken of by the world." — *Gazette Hebdomadaire.*

"Professor Küss's work seems to us to be the best students' manual that we have yet seen." — *Medico-Chirurgical Review.*

"After a careful reading of the book, we do not hesitate to call it, on the whole, the best treatise on Physiology, of its size, now to be found in English. Küss's style is full of vivacity and elegance, and abounds in picturesque epithets and bits of description, which serve both to fix the reader's attention and to impress his memory." — *Boston Medical and Surgical Journal.*

ANATOMY OF THE INVERTEBRATA. By C.

TH. V. SIEBOLD. Translated from the German, with Additions and Notes, by WALDO I. BURNETT, M.D. 1 vol., 8vo, cloth, bevelled boards, gilt top. $5.00.

"The translation is effected in a very satisfactory manner; the language is clear, and conveys the full meaning of the author, without retaining the German idiom. The editor has enriched it with numerous references, and original notes greatly increase the value of the work. We have not the least doubt that the work will speedily supersede every text-book of Comparative Anatomy which has yet appeared." — *Association Medical Journal, London.*

This is believed to be, with the editor's valuable and extensive notes, incomparably the best and most complete work on the subject extant, and as such it is commended by Profs. Agassiz, Silliman, Hitchcock, and other scientific and medical men.

Agassiz says: "There is no other work in any language that will bear any comparison with this, in the fulness and accuracy of its descriptions of organs, in the amount and value of the microscopic investigations whose results it embodies, or in the masterly and comprehensive manner in which all its results are systematized, and their subjects classified and grouped."

THE PROBLEM OF HEALTH AND HOW TO

SOLVE IT. By RUBEN GREENE, M.D. 12mo, cloth. $1.50.

"'The Problem of Health' is a medical work, treating of sanitary science, stimulants, narcotics, sleep, exercise, dress, the value of sunlight, and many other health topics, concerning which a large amount of information is imparted." — *Cape Ann Advertiser, Gloucester, Mass.*

TRANSACTIONS OF THE AMERICAN OTOLOG-

ICAL SOCIETY. Published annually. 8vo, pamphlet. $1.25.

THE DUBLIN PRACTICE OF MIDWIFERY. By

HENRY MAUNSELL, M.D., formerly Professor of Midwifery in the Royal College of Surgeons in Ireland. New edition. Numerous Illustrations. 12mo, cloth. $1.75.

THE HAND-BOOK FOR MIDWIVES. By HENRY

FLY SMITH, B.A., M.B., Oxon., M.R.C.S., Eng., late Surgeon-accoucheur to the St. James Dispensary, late Assistant-Surgeon at the Hospital for Women, Soho Square. 41 Illustrations. 16mo, cloth. $1.75.

THE PASSIONS IN THEIR RELATIONS TO

HEALTH AND DISEASE: Love and Libertinism. Translated from the French of Dr. X. BOURGEOIS, Laureate of the Academy of Medicine of Paris, &c. By HOWARD F. DAMON, A.M., M.D. 16mo, cloth, pp. 224. $1.25.

The following are a few of the notices which have been received:—

"There is a world of suggestions for the management of the passions in this book, and their perusal will not fail to work personal profit." — *Massachusetts Ploughman.*

"There is a delicacy, frankness, candor, and evident sincerity about the composition, that convince even the casual reader that the author and translator have only the welfare of their fellow-men at heart. It is a treatise on Love and Libertinism, in a right, proper, and intelligent spirit, and of incalculable benefit to the whole community." — *The Commonwealth.*

"The book bears no trace of the morbid, unhealthy spirit characteristic of many French books upon this subject." — *Boston Journal.*

COMPARATIVE ANATOMY AND PHYSIOLOGY

OF THE VERTEBRATE ANIMALS. By RICHARD OWEN, F.R.S., Superintendent of the Natural History Departments, British Museum. 3 vols., 8vo, with 1,472 wood-cuts. $25.00.

Vol. I. — *Fishes and Reptiles*, with 452 wood-cuts.

Vol. II. — *Warm-Blooded Vertebrates*, with 406 wood-cuts.

Vol. III. — *Mammalia, including Man*, with copious indexes to the whole work, and 614 wood-cuts.

PHOTOGRAPHS OF THE DISEASES OF THE

SKIN. Taken from Life, under the Superintendence of HOWARD F. DAMON, M D. Photographs, complete (24 Photographs, with letterpress description), quarto, half morocco, $12.00 ; each photograph, without letterpress, 50 cents.

SURGICAL CLINIC OF LA CHARITÉ. Lessons

upon the Diagnosis and Treatment of Surgical Diseases. Delivered in the month of August, 1865, by Prof. VELPEAU. Collected and edited by A. REGNARD, Interne des Hôpitaux. Revised by the Professor. Translated by W. C. B. FIFIELD, M.D. One volume, 16mo, cloth. $1.00.

HAND-BOOK OF THE DISEASES OF THE EYE.

Their Pathology and Treatment. By A. SALOMONS, M.D., Fellow of the Massachusetts Medical Society, former Oculist in Government Service at Veenhuizen, Holland, &c. One volume, 16mo. Colored plate. English cloth. $1.50.

METHOMANIA: A Treatise on Alcoholic Poisoning.
By ALBERT DAY, M.D., Superintendent of the New York State Inebriate
Asylum. One volume, 16mo. Pamphlet, 40 cents; cloth, bevelled boards,
60 cents.

VERATRUM VIRIDE AND VERATRIA: A Con-
tribution to the Physiological Study of. With experiments on Lower Animals, made at La Grange Street Laboratory, 1869. By ROBERT AMORY,
M.D., and S. G. WEBBER, M.D. One volume, 16mo. Pamphlet, 5
cents; cloth, 75.

NITROUS OXIDE: Physiological Action of, as shown
by Experiments on Man and the Lower Animals. By ROBERT AMORY,
M.D., of Longwood, Mass. Illustrated by Pulse Tracings with the Sphygmograph. Pamphlet, 8vo, pp. 31. 50 cents.

TWO CASES OF ŒSOPHAGOTOMY FOR THE
REMOVAL OF FOREIGN BODIES. With a History of the Operation.
Second edition with an additional Case. By DAVID W. CHEEVER, M.D.,
Adjunct Professor of Clinical Surgery at Harvard University, Surgeon to
the Boston City Hospital. One volume, 8vo, cloth. 75 cents.

CONTRIBUTIONS TO DERMATOLOGY. Eczema,
Impetigo, Scabies, Ecthyma, Rupia, Lupus. By SILAS DURKEE, M.D.,
Consulting Physician, Boston City Hospital. Pamphlet, 8vo. $1.50.

PHYSIOLOGICAL AND THERAPEUTICAL AC-
TION AND VALUE OF THE BROMIDE OF POTASSIUM AND THE BROMIDE
OF AMMONIUM. Illustrated by Experiments on Man and Animals.

In Two Parts.

PART I. — The Physiological and Therapeutical Action and Value of the
Bromide of Potassium and its kindred salts. By EDWARD H. CLARKE,
M.D., Professor of Materia Medica in Harvard University.

PART II. — Experiments illustrating the Physiological Action of the Bromide of Potassium and Ammonium on Man and Animals. By ROBERT
AMORY, M.D., Annual Lecturer for 1870-1871 on the Physiological Action of
Drugs in the Medical Department of Harvard University. One volume, 16mo,
cloth. $1.50.

NEW TREATMENT OF VENEREAL DISEASES

AND OF ULCERATIVE SYPHILITIC AFFECTIONS BY IODOFORM. Translated from the French of Dr. A. A. IZARD. By HOWARD F. DAMON, M.D. Pamphlet, 16mo. 50 cents.

THE GYNÆCOLOGICAL RECORD. A Book of

Blank Forms, intended as an aid to the busy practitioner, in recording gynæcological cases, with an Introduction and Appendix of blank leaves, and tables for the ready analysis of the contents of the book. Prepared by JOSEPH G. PINKHAM, A.M., M.D., Corresponding Member of the Gynæcological Society, Fellow of the Massachusetts Medical Society. Approved by the Gynæcological Society. One volume, quarto, half-bound. $2.50. Postage, 50 cents extra. The Blanks, per quire, 50 cents.

PHYSICIAN'S REGISTER, FOR OFFICE OR

HOSPITAL PRACTICE. An Imperial 8vo book of Blank Forms, similar to the book used in the Dispensary, for recording the date, name, residence, age, and disease, with a large blank space for remarks. Price $1.50. 25 cents extra when sent by mail.

HISTORY OF MODERN ANÆSTHETICS. By Sir

JAMES Y. SIMPSON, of Edinburgh. A reply to Dr. JACOB BIGELOW'S second letter. Reprinted from the Journal of the Gynæcological Society of Boston, May, 1870. Pamphlet, 8vo. 25 cents.

THE PHYSIOLOGY OF WOMAN AND HER DIS-

EASES FROM INFANCY TO OLD AGE. Including all those of her critical periods, — Pregnancy and Childbirth, — their causes, symptoms, and appropriate treatment; with hygienic rules for their Prevention, and the Preservation of Female Health. Also, the management of Pregnant and Parturient Women, by which their pains and perils may be greatly obviated. To which is added a Treatise on Womanhood and Manhood, Love, Marriage. and Hereditary Descent; being the most approved views of modern Times. Adapted to the instruction of females. In three books. Complete in one volume. By C. MORRILL, M.D., author of sundry Medical Essays, Lectures on Popular Physiology, &c. Eleventh edition. One volume, 12mo, cloth. $1.50.

THE HISTORY AND PHILOSOPHY OF MAR-

RIAGE; OR, POLYGAMY AND MONOGAMY COMPARED. By a Christian Philanthropist. New and revised Edition. One volume, 16mo, 256 pp. $1.25.

JOURNAL OF THE GYNÆCOLOGICAL SOCIETY

OF BOSTON. A Monthly Journal, devoted to .the Advancement of the Knowledge of the Diseases of Women. Edited by WINSLOW LEWIS, M.D., H. R. STORER, M.D., GEO. H. BIXBY, M.D.

First number was published July, 1869. A few volumes still remain for sale at the prices given below:—

Vol. I. — From July to December, 1869, cloth	$2.50
Vol. II. — From January to July, 1870, cloth	2.50
Vol. III. — From July to December, 1870	2.50
Vol. IV. — From January to July, 1871	2.50
Vol. V. — From July to December, 1871	2.50
Vol. VI. — From January to July, 1872	2.50
Vol. VII. — From July to December, 1872	2.50

DISEASES OF THE WOMB. Uterine Catarrh fre-

quently the Cause of Sterility. New Treatment By H. E. GANTILLON, M.D. Pamphlet, 8vo. 50 cents.

"This little brochure is well worthy the study of all who are interested in Gynæcology." — *St. Louis Medical and Surgical Journal.*

THE DETECTION OF CRIMINAL ABORTION,

AND THE STUDY OF FŒTICIDAL DRUGS. By ELY VAN DE WARKER, M D. Illustrated by Pulse Tracings with the Sphygmograph. Pamphlet, 8vo. 50 cents.

"It is a very sensible and thorough treatise on this important subject, and should be read by the profession everywhere " — *Boston Journal of Chemistry.*

THYROTOMY FOR THE REMOVAL OF LARYN-

GEAL GROWTHS. Modified. By EPHRAIM CUTTER, M D. Illustrated. Pamphlet, 8vo. 50 cents.

FEMALE HYGIENE. A Lecture delivered at Sacra-

mento and San Francisco. By HORATIO R. STORER, M.D. Pamphlet, 8vo. 25 cents.

"It is not only an admirable treatise on a subject on which the author is especially qualified to write, but it also does good service in combating the woman suffrage delusion." — *Boston Traveller, March, 18, 1872.*

The Publisher would invite the attention of the public to the following admirable reviews of Dr. Simon's book on Filth Diseases, *and would urge upon every one the importance of a careful examination of the book itself.*

From the Boston Traveller.

It is comparatively rare that a work by a thoroughly scientific medical man comes from the press in such a shape as to be of practical value to the non-professional reader. Either from the terminology employed, or the subject treated, medical books and reports are only to be found on the shelves of the physician's library. Such is not, or should not, be the case with a small volume recently published by James Campbell, of this city, it being the American reprint of a most masterly essay by John Simon, chief medical officer of the Privy Council and of the Local Government Board of England. The essay is most cordially commended by the members of the State Board of Health of Massachusetts, and certainly no intelligent householder should fail to read it carefully, and profit by the information contained in its pages. Its title, "Filth Diseases and their Prevention," indicates the particular line of investigation followed by Dr. Simon; and his conclusions, based on statistics and results of careful investigation, are thoroughly logical. He assumes at starting that the *raison d'être* of sanitary authorities, like our boards of health, is the fact that very much disease is preventable; and that it is true that the mortality from diseases is vastly greater than it would be if the existing knowledge of the causes of disease were applied. Of all the removable causes of disease, Dr. Simon justly considers the chief to be uncleanliness: that is, first, the non-removal of refuse matters; and, second, the license permitted to cases of infectious disease to scatter the seeds of infection. He says that a bad odor is by no means a sure warning against the presence of poisonous matters. That they may exist without any odor whatever, and that disinfection by no means consists in covering up one bad smell by another equally offensive but more pronounced. He goes thoroughly into the subject of disinfection, and shows just how it should be done to be of any value. Interesting cases are quoted, showing in what subtle ways these *ferment poisons*, such as cause typhoid fever, are spread abroad, manifesting their results miles and miles away from their source, being carried in air, water, milk, and other vehicles suitable to preserve their vitality. The subjects of typhoid fever and cholera are quite fully discussed in relation to their preventability, as well as the relation of cause and effect which filth may bear to consumption.

A large portion of the essay is devoted to the question of house drainage and public sewerage, with suggestions of the utmost importance to every householder. Dr. Simon shows just how and in what particular way the public sewers, when insufficient or defectively ventilated, may become exceedingly dangerous. Apropos of this subject of ventilation of sewers, it is interesting to see how his judgment coincides with that of the gentlemen who opened the rain pipes into the sewers, in 1874, in this city, and for which they have been soundly abused by some medical men, particularly by one recently in the columns of a morning paper. That writer made out a frightful increase in mortality by comparing statistics of 1874 and 1875, when, as he admits, the rain pipes were let into the sewers during both years. He, however, unfortunately for his argument, took the total mortality instead of the mortality from zymotic diseases. Now, he can hardly claim that sewer gas causes apoplexy or heart disease; and, if he takes only the diseases which can be claimed to be caused by filth, he will find a very gratifying decrease in the mortality in the first three months of 1875 when the sewers were ventilated, as compared with the same months in 1874 when they were not It is simply because the sewers are not yet sufficiently ventilated that we occasionally notice the offensive odors. Dr. Simon considers this matter of ventilation of sewers of the utmost importance. His essay is delightfully clear, and free from technical terms, and can be read with pleasure and profit by every person of ordinary intelligence; and, if landlords will act on his suggestions, much sickness and death may be prevented.

JUST PUBLISHED:

A NEW MANUAL OF PHYSIOLOGY.

A Course of Lectures on Physiology,

As delivered by Professor Küss at the Medical School of the University of Strasbourg. Edited by MATHIAS DUVAL, M.D., formerly Demonstrator of Anatomy at the Medical School of Strasbourg. Translated from the Second and Revised Edition, by ROBERT AMORY, M.D., formerly Professor of Physiology at the Medical School of Maine. 150 Woodcuts inserted in the text. 1 vol. 12mo. 547 pp. Price $2.50.

"M. Küss was one of those modest scientists who love science for itself; who seek for the truth without caring whether they are well spoken of by the world." — *Gazette Hebdomadaire.*

"The author exhibits a thorough familiarity with the late advances made in physiological science; and, although we have a number of acceptable works on this subject, we welcome this as one particularly well adapted to advanced students. Its terseness gives the reader and student an impression that it is really a great and large work, boiled down to the dimensions of a handbook." — *Cincinnati Lancet and Observer.*

"I have a good many works on the subject, but all of them seem to me in some respects a little antiquated; and, in the examination I have made of these lectures, they seem to me to meet just the want which I and others of my time feel very urgently. I am also pleased to have some new illustrations, after meeting the old stereotyped ones so many times." — *Professor Oliver Wendell Holmes.*

"The arrangement of this manual of Physiology is judicious, and its discussions of the various subjects involved concise and accurate." — *Philadelphia Medical and Surgical Reporter.*

"After a careful reading of the book, we do not hesitate to call it, on the whole, the best treatise on Physiology, of its size, now to be found in English. Küss's style is full of vivacity and elegance, and abounds in picturesque epithets and bits of description, which serve both to fix the reader's attention and to impress his memory." — *Boston Medical and Surgical Journal.*

"This manual is the only concise treatise wherein the relations of Physiology to Histology are carefully presented, in the English language. The illustrations are both numerous and well executed." — *Physician and Pharmacist.*

For sale by all Booksellers, or forwarded by mail to any part of the United States on receipt of the price and twenty-five cents extra, to prepay postage, by

JAMES CAMPBELL,

PUBLISHER, BOOKSELLER, AND STATIONER,

18 TREMONT STREET, BOSTON, MASS.

SURGICAL CLINIC OF LA CHARITÉ.

LESSONS

UPON THE

DIAGNOSIS AND TREATMENT OF SURGICAL DISEASES,

Delivered in the month of August, 1865, *by Prof. Velpeau.*

COLLECTED AND EDITED BY A. REGNARD, INTERNE DES HOS-
PITAUX. REVISED BY THE PROFESSOR.

TRANSLATED BY W. C. B. FIFIELD, M.D.

1 volume. 16mo. Cloth. $1.00.

NOTICES OF THE PRESS.

"This modest little book contains a statistical *résumé*, by the author, of his surgical experience in the hospital wards under his care during the year. He treats his
subject under the successive headings: Generalities, Fractures, Affections of the
Joints, Inflammation and Abscesses, Affections of the Lymphatic System, Burns and
Contusions, Affections of the Genito-Urinary Organs, Affections of the Aural Region,
Affections of the Eyes, Statistics of Operations. We have a special liking for such
works, which give us the most authoritative opinions of the elders of the medical profession, who have reached the time when the judgment is least biased by the rivalries
and personal influences which are so apt to mislead younger minds. It is o vastly
more value than many more ambitious and bulky works." — *Boston Medical and
Surgical Journal.*

" He not unfrequently surprises us by the simplicity of his expedients for the aid
of ' Nature in Disease,' and rarely, if ever, fails in making out his case. As a whole,
the work is not only instructive, but entertaining, and may be regarded as one of our
landmarks of minor surgery, upon our skill in which much of our success will be
found to depend." — *Medical Record.*

" It is rare that so small a book contains so many suggestions of great practical
worth, and throws so much light on certain debated points, as Velpeau's Lessons.
Though nominally a review of one year's practice, it is in reality an epitome of the
experience of a lifetime." — *Detroit Review.*

"All who value the teachings of this great man will not lose the opportunity of
obtaining them, as presented in this brief and economical form." — *Richmond Medical Journal.*

Sent by mail, postage prepaid, on receipt of advertised price.

JAMES CAMPBELL, PUBLISHER,

18 *Tremont Street, Museum Building, Boston, Mass.*

HAND BOOK

OF THE

DISEASES OF THE EYE.

Their Pathology and Treatment.

BY

A. SALOMONS, M.D.,

Fellow of the Massachusetts Medical Society; former Oculist in Government Service
at Veehnhizen, Holland, &c.

One volume, 16mo. Colored plate. English cloth. $1.50.

---◆---

PREFACE.

" The book is divided into two parts: the first includes the
pathology and treatment of eye diseases; and the second, the
operative surgery of the eye. The practical portions of the work
are given with as much detail as possible, and from the experience of the author; and it is hoped they may prove a useful
guide, not only to those entering this interesting department
of medicine, but also to the busy practitioner, who finds himself
unable to peruse the more elaborate treatises on this subject."

From the Philadelphia Medical and Surgical Reporter.

" A synopsis like this, which goes over so much ground in so small a space, is
advantageous to the student, in connection with clinical studies, and the perusal of
more extended treatises. The definitions are carefully given, accuracy is observed,
and lucidity is not sacrificed to brevity. The operations recommended are carefully
selected and described. That for Entropium we may particularly mention as in point."

---◆---

*For sale by all medical booksellers, or sent by mail, postage
prepaid, on receipt of advertised price.*

JAMES CAMPBELL, PUBLISHER,

Boston, Mass